How To Build And Frame Winder Stairs

By Greg Vanden Berge

Published by Greg Vanden Berge
Copyright 2015 Greg Vanden Berge

ISBN-13: 978-1514224779
ISBN-10: 1514224771

Greg's Other Books

Home Buyers Checklist
How To Build Straight Stairs
501 Contractor Tips
Simplified Stair Building
Guide For Hiring Contractors
Simplified Bracket Stair Building
Simplified Tile Floor Installation
Simplified House Inspection Checklist
Simplified Home Inspections
Advanced Stair Stringer Layout Methods

Author's Website: http://gregvandenberge.com

Construction Books: http://gregvan.com/book_deals.htm

Stair Building Website: http://stairs4u.com

Disclaimer

Greg Vanden Berge, and its owners, agents and employees, make no warranty respecting the accuracy or currency of any information in the content or pages of this book or any source document referenced herein or linked to herein. Use of this book is conditioned on the user's understanding and agreement that we shall not be liable, on any theory whatsoever, including but not limited to negligence, for any damages attributable to that use.

In no event shall Greg Vanden Berge, its owners, agents or employees be liable to you or anyone else for any decision made or action taken by in reliance on any content created by Greg Vanden Berge or other individuals, companies, corporations or parties.

Greg Vanden Berge and its affiliates, agents, owners and employees shall not be liable to you or anyone else for any damages, including without limitation, consequential, special, incidental, indirect, or similar damages, even if advised of the possibility of such damages.

Your use of this book and all related rights and obligations, shall be governed by the laws of the United States of America, as if your use was a contract wholly entered into and wholly performed within the United States of America.

Any legal action or proceeding with respect to this book or any matter related thereto may be brought exclusively in the courts of the United States of America. By using this book, you agree generally and unconditionally to the jurisdiction of the aforesaid courts and irrevocably waive any objection to such jurisdiction and venue.

Do not copy or distribute this book. This manual contains materials protected under International and Federal Copyright laws and Treaties. Any unauthorized reprint or use of this material is prohibited.

Table of Contents

Introduction	4
What is a Stairway	5
Tools	6
Hardware and Fasteners	7
Stair Parts	8
Safety	9
Engineering	10
Calculating Riser Height	13
Calculating Tread Run	16
17.5 Tread and Riser Rule	18
Layout and Measurements	20
Stair Stringer Layout	22
Installing Stair Stringers	45
Laying Out Treads and Risers	48
Installing Risers	49
Installing Treads	51
Glossary	55
Rise and Run Chart for 10 Inch Treads	63
Rise and Run Chart for 17-½ Rule	77
Decimals to Inches Chart	91
Winder Basics	92
Winder Examples	100
How to Layout Winder Treads	135
Build it in Sections	142

Introduction

This book is part of a series designed for professionals and do-it-yourselfers to provide them with what I consider to be a simplified step-by-step process for designing and assembling different types of stairs. Each book will be written and illustrated specifically for the type of stairway specified in the title.

Book 1 - How to Frame and Build Stairs

This book provides you with step-by-step detailed instructions on how to design, layout stair stringers and build a variety of different sized straight stairways. Sections of this book or the entire book will be included in some of the other books as noted.

Book 3 - How to Frame and Build Winder Stairs

This book includes book 1 and will provide you with step-by-step detailed instructions on how to design, position and build winder stairs.

This is the first series of stair building books written by Greg Vanden Berge, but they are not the first books he has written about stairs. For more information or to contact us you can visit the website: **http://stairs4u.com**.

Repeat... Do Not Purchase Book 1 Because It Is Included In This Book!!!

What is a Stairway?

A stairway or set of stairs is usually a series of steps used to take you from a lower level to an upper-level.

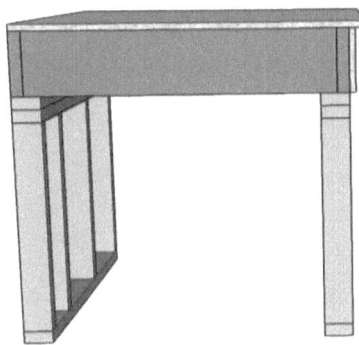

Before

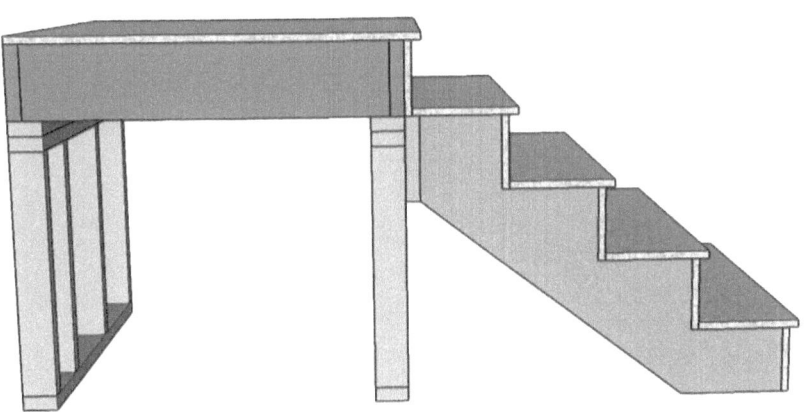

After

Tools

Hammer

Level

Pencil

Circular Saw

Framing Square

Tape Measure

Caulk Line

Optional Tools – nail puller, ladder, carpenter pouch or nail bags.

Hardware and Fasteners

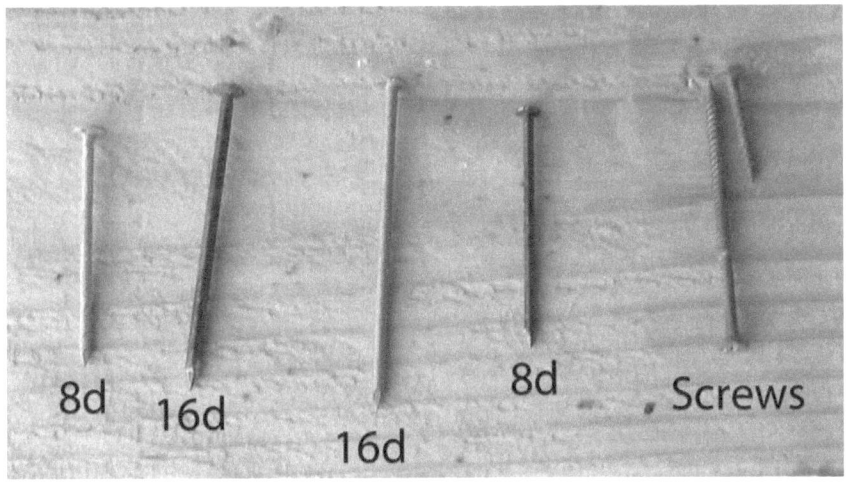

Recommend using 16d nails for materials between an inch and one quarter (1 1/4") and an inch and a half (1 1/2").

Recommend using 8d nails for materials between one half-inch (1/2") and three quarters of an inch (3/4").

I can't recommend screw sizes, because some building departments and engineers don't recommend them to be used for assembling stairs using construction standard lumber.

I also recommend using galvanized or stainless steel nails for building "exterior" stairs out of wood.

For additional holding power you can use ring shank, hot dipped or drive screw deformed shank type nails for stair treads and sheathing.

Stair Parts

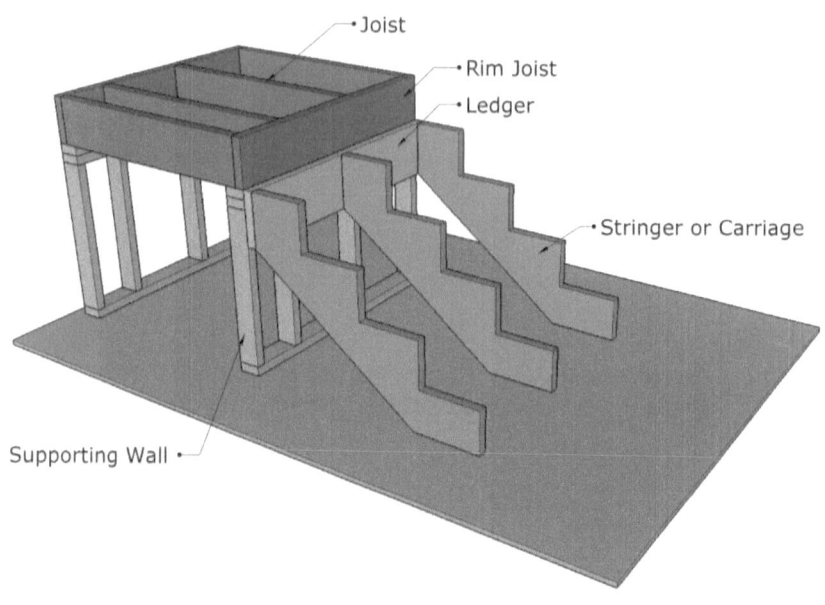

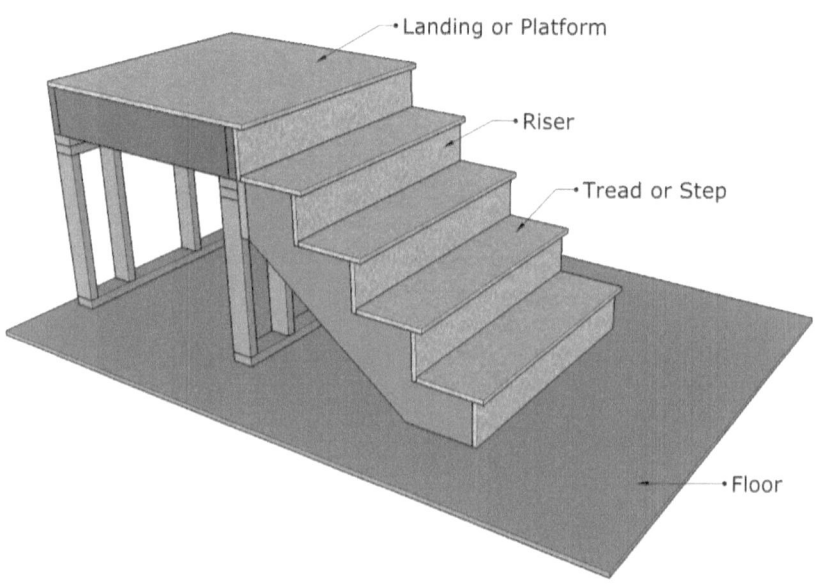

Safety

This book does not provide you with detailed safety information and recommends gathering more information on tool use, demolition, safety and assembling practices.

Safety Recommendation Tips

1. Avoid standing on stair stringers, joist and other framing members that haven't been properly secured.

2. Avoid using power tools near water and make sure they are in good condition.

3. Use personal protection equipment like safety glasses, slip resistant shoes, earplugs, hardhats and gloves when necessary.

4. Inspect all ladders for defects before using them and make sure you use the appropriate ladder for each phase of the project.

5. Use extra caution when working with any materials, including prefinished materials like stair treads that are or can become slippery.

6. Keep the job site and working areas clean.

7. Be prepared for emergencies with an on-site first aid kit and directions to the nearest hospital or emergency care facility.

8. Always use the right tool for the right job.

9. It's a good idea to create a barricade around the area you're going to be working in, especially when working on your own home around pets and children.

10. Avoid lifting heavy objects and work smart.

Engineering

I won't be able to provide you with exact lumber sizes for your particular project, but can provide you with a few recommended lumber sizes from projects I've built in the past that were approved by building professional's.

Again, these are only recommended construction standard lumber sizes and might not work on your project.

Estimated Stair Stringer Sizes

2 x 12 for lengths up to 12 feet.

2 x 14 for lengths up to 18 feet.

You would need to check with the manufacturer for stringer sizes if you're planning on using engineered lumber.

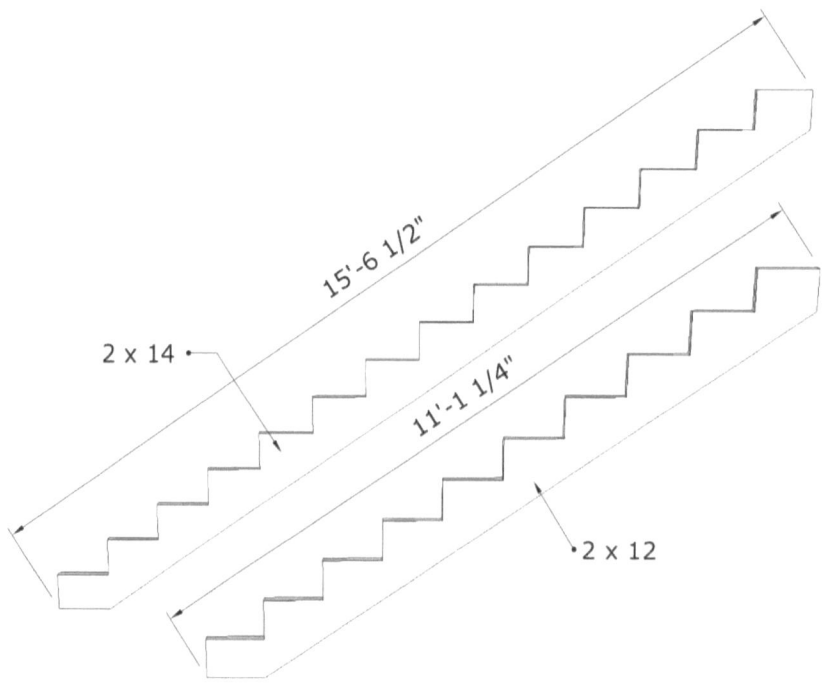

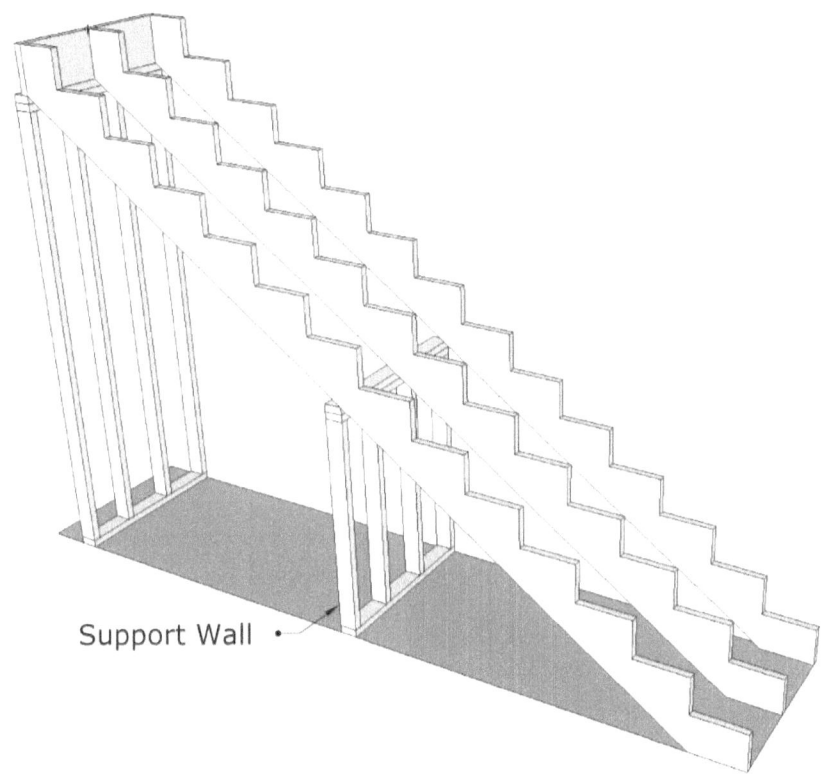

Support Wall

Stair stringers longer than 10 feet might require supporting walls. 2 x 4 can usually be used to build supporting walls less than 10 feet high for stringers, landings and floors.

Landing or Platform Joist

2 x 6 joists usually works for most landings with joist spans less than 6 feet and spaced 16 inches on center.

2 x 8 joists usually works for most landings with joist spans less than 8 feet and spaced 16 inches on center.

2 x 10 joists usually works for most landings with joist spans less than 12 feet and spaced 16 inches on center.

Use metal building hardware or treated lumber framing base plates when attaching a wood framed stairway to concrete.

Stair Treads and Stringer Spacing

Stair treads and risers can be made out of engineered lumber like plywood and oriented strand board or construction standard lumber.

Estimated stringer spacing for ¾" engineered lumber used for treads would be no more than 20 inches.

Estimated stringer spacing for 1 - 1/8" engineered lumber used for treads would be no more than 24 inches.

Estimated stringer spacing for 2 x 6 construction standard nominal sized lumber used for treads would be no more than 20 inches.

Estimated stringer spacing for 2 x 12 construction standard nominal sized lumber used for treads would be no more than 24 inches.

Estimated stringer spacing for 3 x 12 construction standard nominal sized lumbers used for treads would be no more than 36 inches.

Estimated stringer spacing for 4 x 12 construction standard nominal sized lumbers used for treads would be no more than 48 inches.

The lumber sizes I've just provided you with are fairly conservative and should work with most engineering load requirements of 40 pounds per square foot or less which is usually acceptable for residential stairways.

These are only estimates and might not work for stairways designed for the public, commercial or industrial type buildings. These buildings usually require larger lumber sizes to handle extra weight from additional people and large items.

Calculating Riser Height

If you have a set of building plans then you should already have the correct individual tread and riser measurements. This chapter and the one that follows will be more helpful to those who will be designing their own stairs.

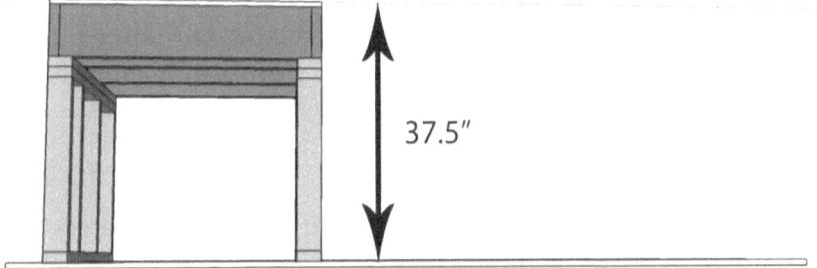

The first thing we need to do is measure the total stair rise which is the measurement between the lower and upper level. In our example we will be using 37 ½ inches and will provide you with a simple and easy formula for figuring out your individual stair riser measurement in the next two pages.

The next thing we need to do is figure out how many stair steps (treads) and risers we will need for the stairway and the easiest way I can think of doing this will be to divide the stair rise by the number seven.

Example: 37.5 ÷ 7 = 5.36 and then we will round the number off to its nearest whole number (5). The number five will provide us with a good place to start, but it won't be a bad idea to divide the next smaller and larger whole numbers into the total rise to find the most comfortable individual riser measurement.

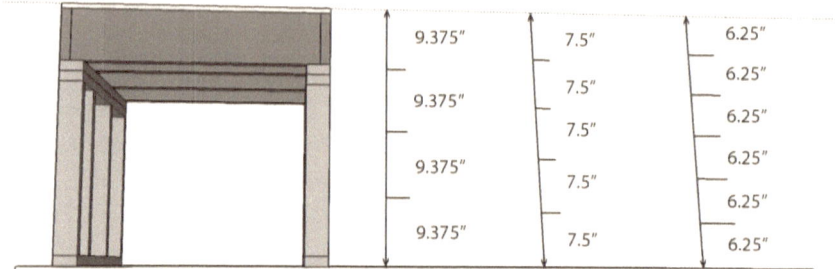

In the example above we will divide 4 (the next smaller whole number), 5 and 6 (the next larger whole number) into the total rise of 37.5 inches.

37.5 ÷ 4 = 9.375″   37.5 ÷ 5 = 7.5″   37.5 ÷ 6 = 6.25″

It provides us with two choices, because some building codes don't allow risers to be taller than 7 ¾″ or smaller than 4 inches, leaving us to choose from either a 7.5 or 6.25 inch measurement.

Either one is acceptable, but before making your final decision, I suggest reading the next two chapters especially the one on the 17 ½ inch rule.

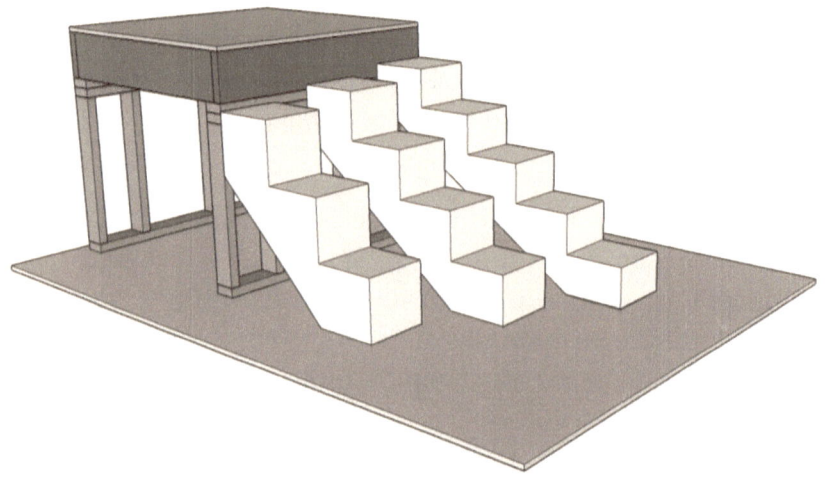

Calculating Riser Height Formula

Using Previous Example

Total Rise in Meters or Inches _ 37.5

Total Rise _ 37.5   Divided by Seven Equals _ 5.36

Rounded off Whole Number (B) _ 5

Next Smallest Number (A)   _ 4

Next Largest Number   (C)   _ 6

Total Rise_37.5   Divided (A) _4   Equals _9.375

Total Rise_37.5   Divided (A) _5   Equals _7.5

Total Rise_37.5   Divided (A) _6   Equals _6.25

You Can Use Either Meters or Inches in Formula Below.

Total Rise _____

Total Rise_____ Divided by Seven Equals _____

Rounded off Whole Number (B) _____

Next Smallest Number (A) _____

Next Largest Number (C) _____

Total Rise_____ Divided (A) _____ Equals _____

Total Rise_____ Divided (B) _____ Equals _____

Total Rise_____ Divided (C) _____ Equals _____

Calculating Tread Run

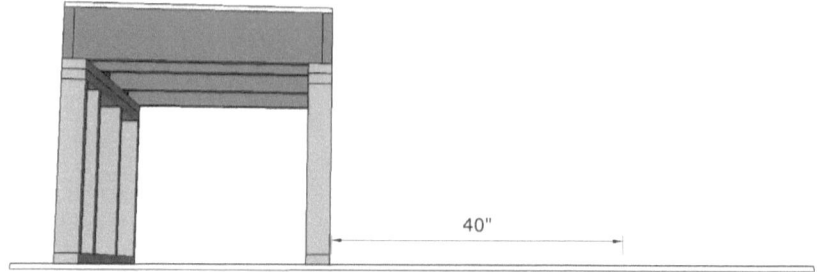

We will use the same method we used to calculate our risers. Simply take the measurement you're going to use for the overall tread run and divide it into equal measurements.

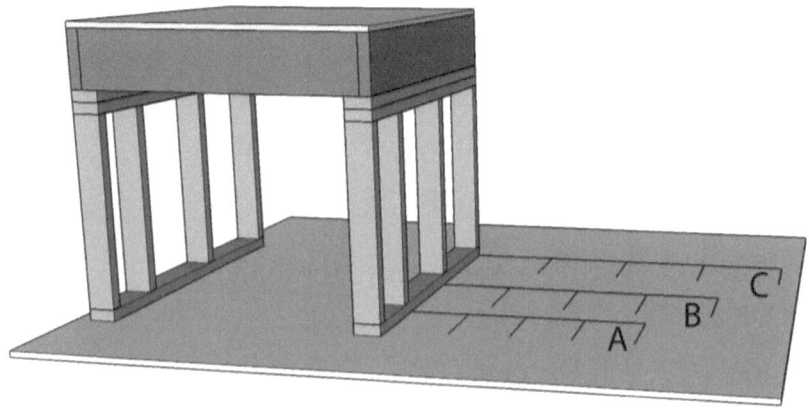

A - Tread Run = 32 inches and four 8 inch treads.

B - Tread Run = 40 inches and four 10 inch treads.

C - Tread Run = 48 inches and four 12 inch treads.

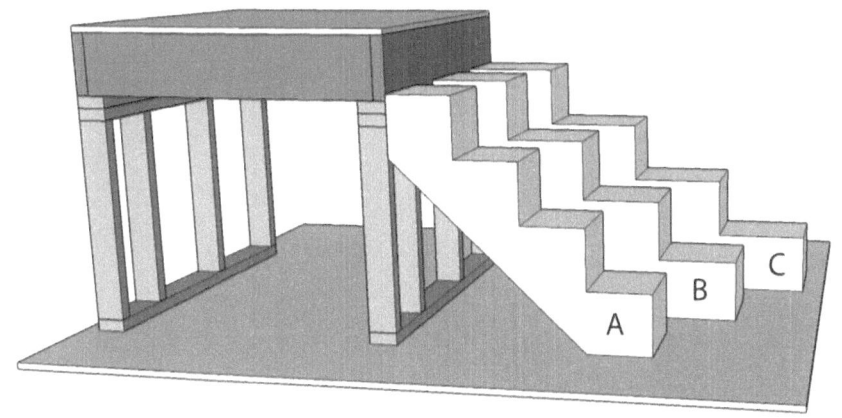

There is always going to be one more riser than treads, so you can use this to figure out either the amount of treads or risers you'll be using for your stairway.

I suggest figuring the amount of risers that will be used and then subtracting one from the total amount to calculate the amount of treads.

For Example: If you have 14 individual 7 ¾" risers then you will have 13 individual treads. It's hard to do it the other way around.

For Example: If we use the same situation as the example above we would have a total rise of 108 ½". If we decided to use 11 individual treads and added one to come up with the total number of individual risers it would give us 12. If we divide 12 into 108 ½" we would end up with a 9 inch individual riser height.

There will be times when you need to start with the total tread run, but most of the time starting with the total rise will provide you with the best way to figure out the total run, amount of treads and individual tread width.

17.5 Tread and Riser Rule

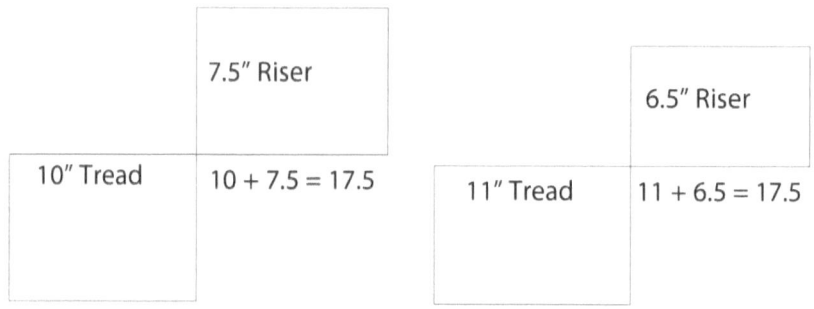

Here's a method I've been using for years to provide the most comfortable step possible, even though you might not be able to use it all the time, because it won't work in confined areas.

Simply add the individual tread and riser measurements together as shown in illustrations above. The closer you come to 17 ½ inches the better.

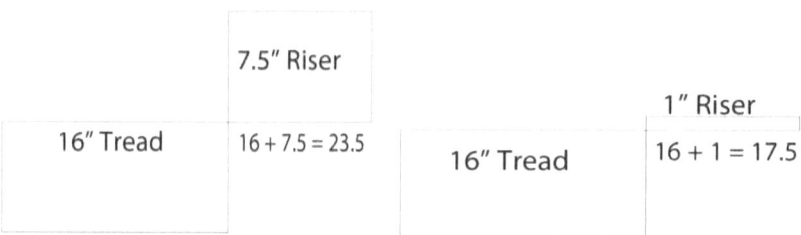

In the example above we are way off with a 16 inch tread and 7 ½ inch riser, but the solution to the problem wouldn't be to use a 1 inch riser. It would be better to use a smaller tread or a smaller riser as long as the total individual riser height isn't less than 4 inches.

Remember, some residential building codes discourage using risers that have an individual height of more than 7 3/4" or a minimum height less than 4 inches and I agree.

The example above also doesn't provide us with something reasonable, even though the numbers add up to 17 ½ inches.

I don't suggest using individual stair tread widths less than 10 inches and some building codes don't recommend using treads less than 11 inches. As you can see by the illustration in the upper right-hand corner, people with larger shoes might have problems using stairs with smaller tread widths.

Remember, you don't need to build a stairway using this formula, but it does provide us with a comfortable step if used within reason.

Listed below are a few examples of what I would consider to be comfortable steps.

10 inch stair tread width + 7 to 7-3/4" riser height.

10-1/2 inch stair tread width + 6-1/2 to 7-1/4" riser height.

11 inch stair tread width + 6 to 6-3/4" riser height.

11-1/2 inch stair tread width + 5-1/2 to 6-1/4" riser height.

Even though 17 ½ inches will provide us with a comfortable step, anything between 17 and 18 inches should be acceptable. Anything between 16 ½ and 18 ½ inches could be acceptable also, but once you start to get out of this range, then steps might start to get uncomfortable.

Layout and Measurements

By now you should have figured out the overall height (total rise) and the overall length (total run) along with the individual riser height and tread width. However, there are a few things I would like you to check before laying out your stair stringer.

1. Is there anything you need to add or subtract to the total rise or run and if so then you will need to recalculate to find the new individual tread or riser measurements?

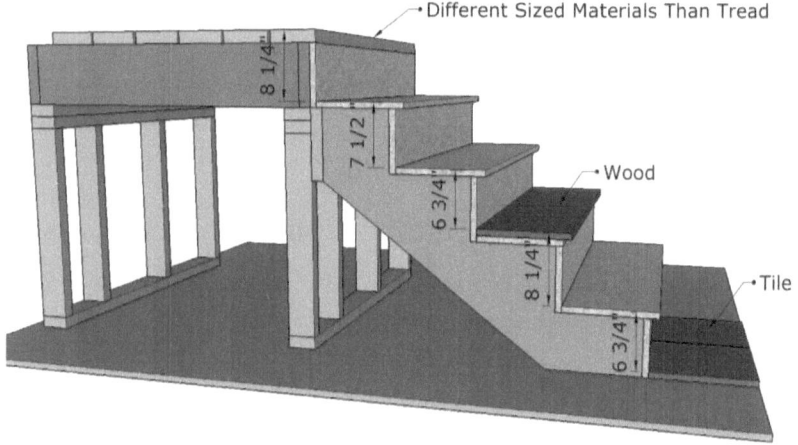

If you're going to install different sized materials on either the stairway steps or upper and lower levels like floors or landings then you will need to make the necessary adjustments while laying out the stair stringer.

2. Is it going to fit into the designated area?

Now would be a good time to double check all of the stairway measurements by re-measuring the total rise and total run and verifying that you have done all calculations correctly.

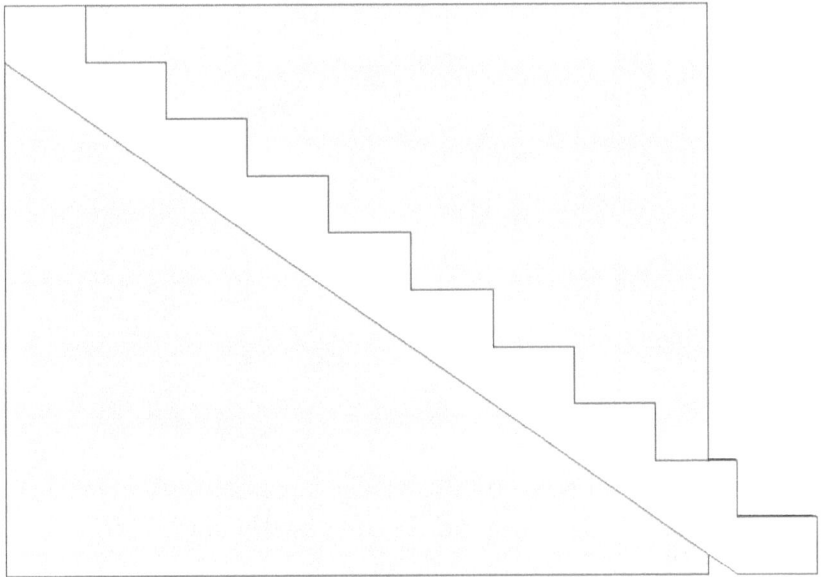

3. Is everything square and parallel?

The area where the stairway is going to be installed is usually called the stairwell and parallel walls will need to have the same measurement at both ends and in the middle. 90° perpendicular walls will need to be square or as square as possible in the stairwell.

You should also check to make sure all connecting walls are level and plumb. It will be easier to make the necessary adjustments before the stairway is installed.

Stair Stringer Layout

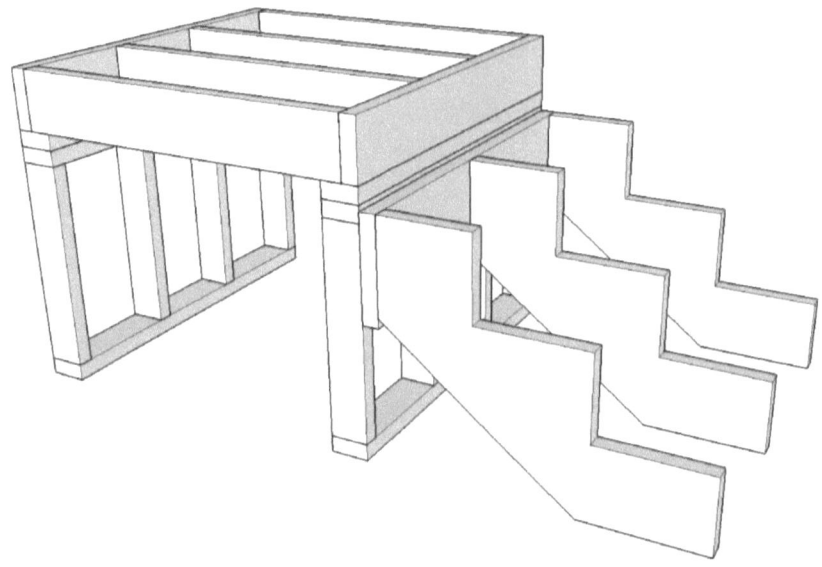

For those of you who have never built a set of stairs, it wouldn't be a bad idea to read the rest of the book before laying out your stair stringer pattern.

If you end up reading the rest of the book but don't quite understand what you're doing then feel free to start at the beginning of the book and work your way through the process step by step.

Some people learn better by reading, while others need to actually do it. I strongly suggest double checking and even triple checking all measurements you make throughout the entire planning, layout and assembly of the stairway.

You can also visit the website for how to videos.
http://www.video.stairs4u.com

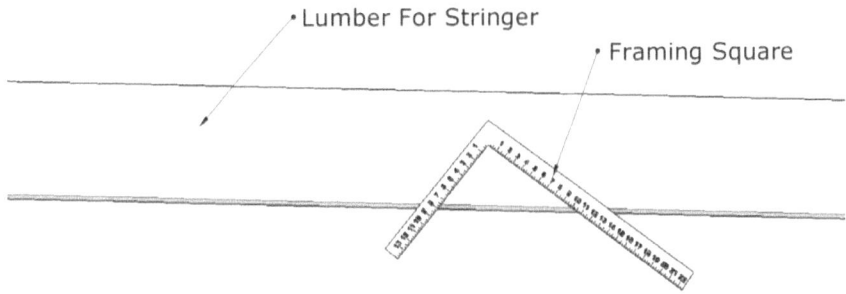

Place the lumber you're going to use for your stair stringer pattern on top of a couple of saw horses or table to make it easier to work. I like to use 2 x 12 or 2 x 14 construction standard lumber with a lumber grade of number one or better if available for stringers.

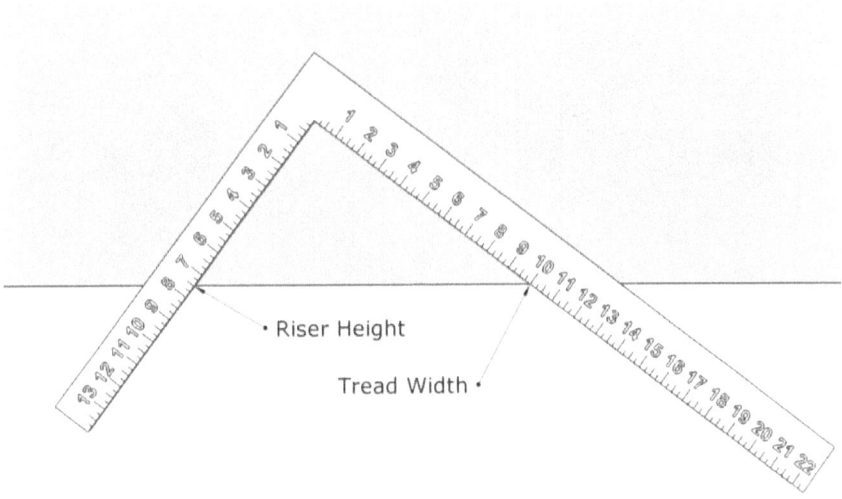

Place the framing square on top of the lumber and line up the marks you're going to use for the tread width and riser height.

In our example we're going to use a 10 inch tread width, even though your measurement might be different. Remember to line "Your Measurements" up with the corner edge of the stair stringer as shown in illustration above and not ours.

Do the same with the riser height measurement on the other side of the framing square. Again, the riser height measurement might be different so use "Your Measurements", unless they are the same.

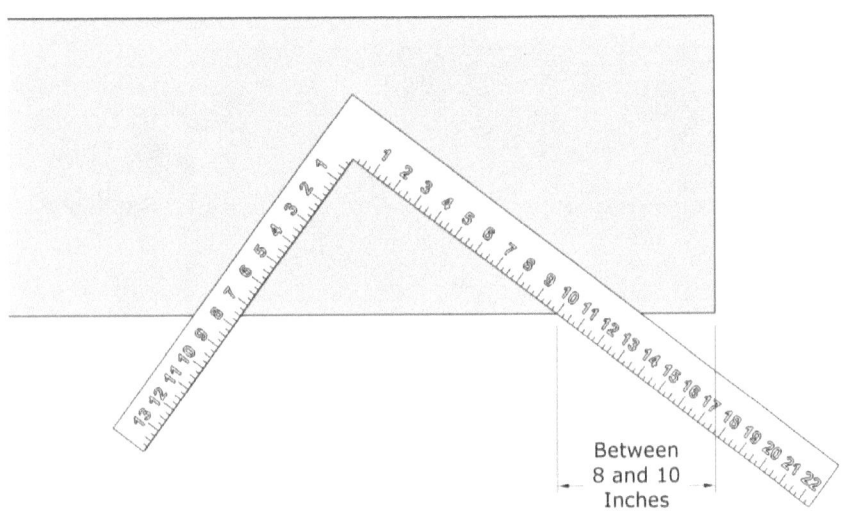

Between 8 and 10 Inches

I like to hold the framing square between 8 and 10 inches away from the edge as a starting point. However, if there are cracks or large knots then this measurement will need to be longer.

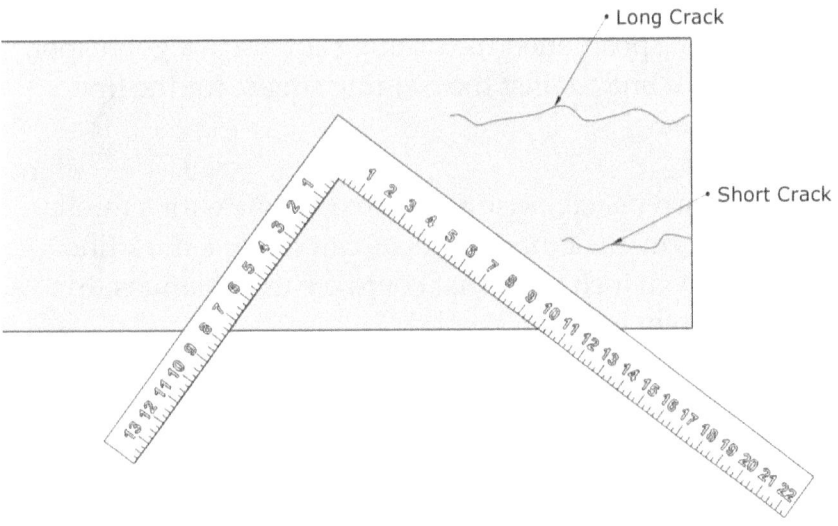

Long Crack

Short Crack

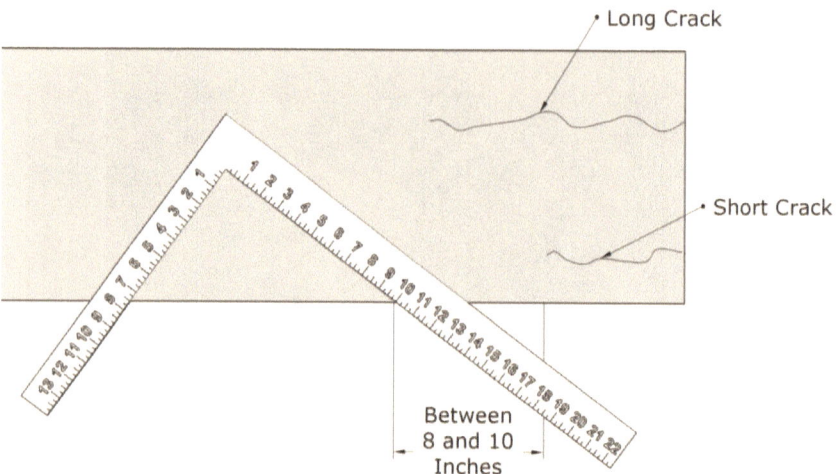

Simply move the framing square over or select a different piece of lumber for your stair stringer pattern.

Before drawing the line, make sure everything is lined up perfectly and you can use a pencil, pen or permanent marker to draw each line.

After you have positioned the framing square in a good spot, simply draw a line against the framing square for the first tread and riser.

I prefer a sharp pencil, because it provides me with a thinner line. Some permanent marking pens can create a thick line almost 1/8 of an inch wide that could create variations up to a 1/4 of an inch in between steps.

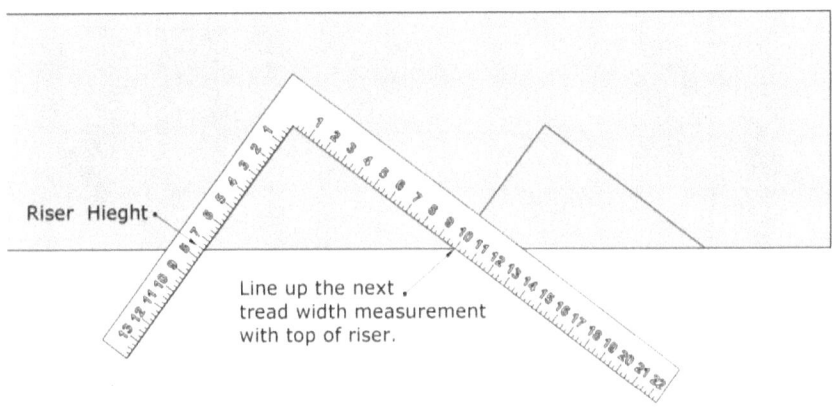

Besides figuring your riser height and tread width correctly, this next step will probably be the most important part of your stair building project.

As shown in the illustration above, you will need to line up the riser height and tread width measurements exactly the same way you did in the previous step, except this time you will align the tread width measurement with the top edge of the riser height measurement.

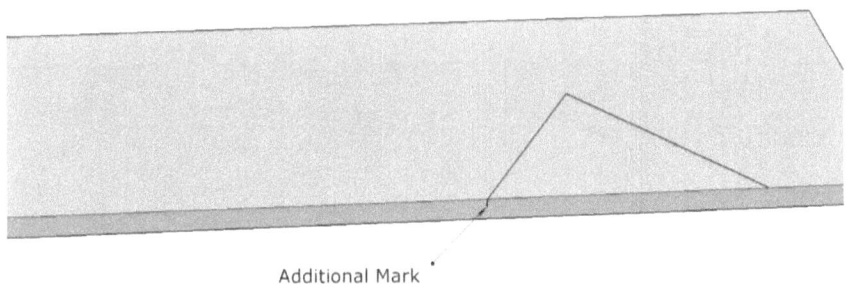

Additional Mark

If you have a problem finding the exact spot, you can always create an additional mark on the edge for better viewing.

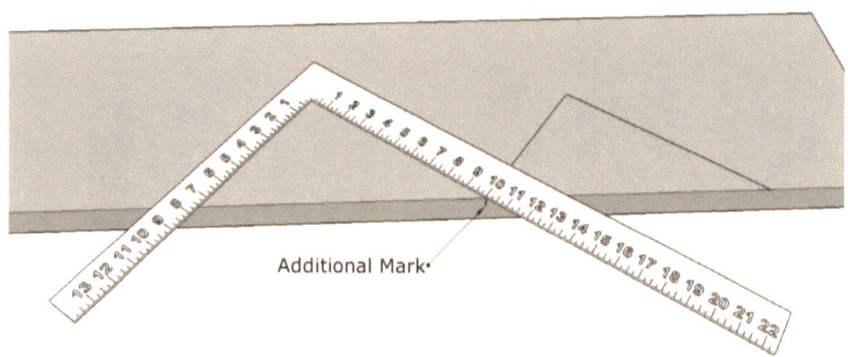

After you have lined everything up in the correct position, go ahead and mark the next riser and tread.

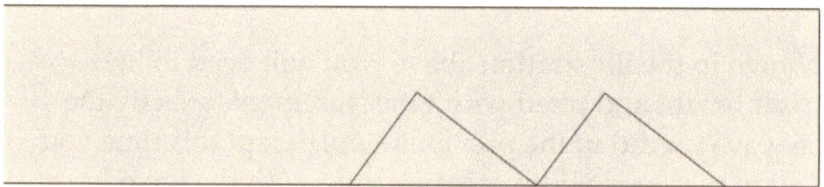

By now you should have something that looks like this.

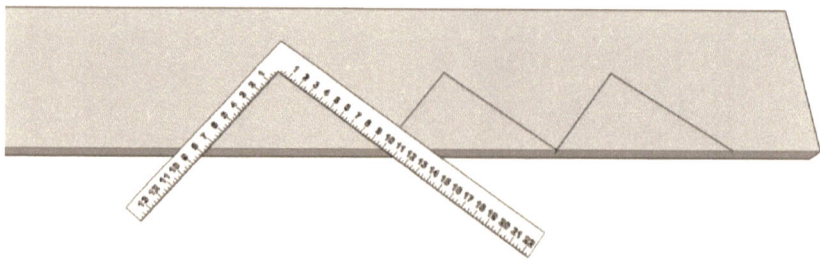

Repeat this process until you have laid out all of your treads and risers.

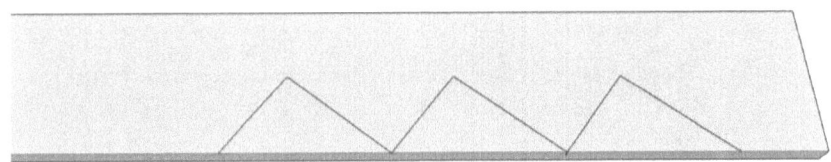

In our stair stringer example, we only needed three steps, so we're going to stop here.

However, this might not be the case for your stair stringer. If you have 10 risers and 9 treads then you will need to continue working your way down the stair stringer pattern until you have made one riser height and tread width mark for each stair step or tread.

In the next two sections I will show you how to layout the top and bottom of your stair stringer. Keep in mind there are different ways to layout the tops and bottoms of stair stringers, but these are the most common and easiest to understand.

For more information, check out our other books by visiting the link below.

Advanced Stair Stringer Layout Methods

Professional Stairway Building Secrets

http://stairs4u.com/stairbuildingbooks.htm

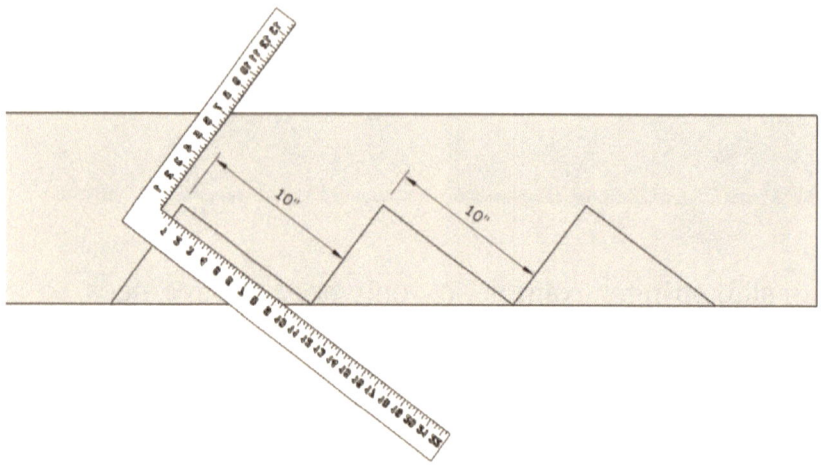

Double check your stair tread width measurement and riser heights before continuing. If there are any variations greater than one quarter of an inch, then you should get rid of the pattern and start over.

For example, if you have one 10 inch tread and the next one is 10 1/4 inches or you have a riser height of 7 1/2 inches on your second riser and 7 1/4 inches on your third one, then you should figure out what you did wrong and layout a new pattern.

I personally wouldn't allow more than an eighth of an inch, yet most building codes allow up to 3/8 of an inch. Something like this might not be a problem on stairs with less than four steps, but could create an extremely unsafe stairway to walk up and down on any stairway with more than six steps.

For more information on this building code, visit our website.

http://stairs4u.com/code/maximum_variation_in_stair_riser.htm

Laying Out Top

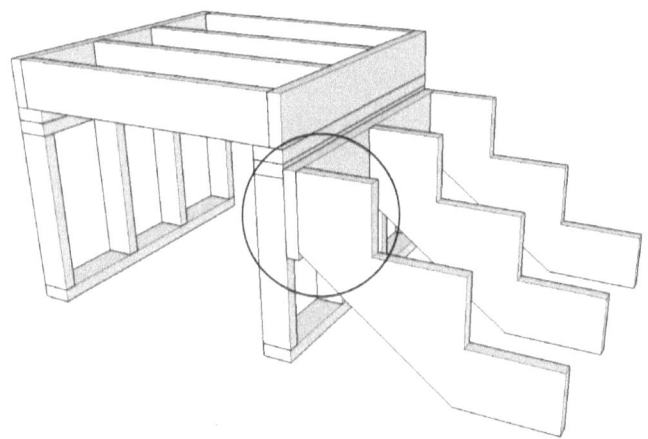

This layout method will require a stair stringer ledger and supporting wall. The ledger will either be a 2 x 10 or 2 x 12 and will attach to a wall. Other materials can be substituted, but for our example we are going to use a 2 x 10 that's 1-1/2 inches thick by 9 1/2 inches in height.

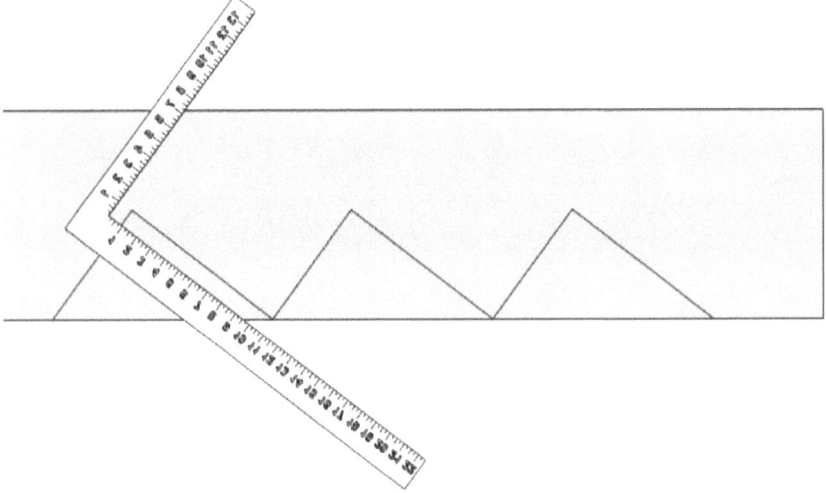

Rotate and place the framing square as shown in picture above.

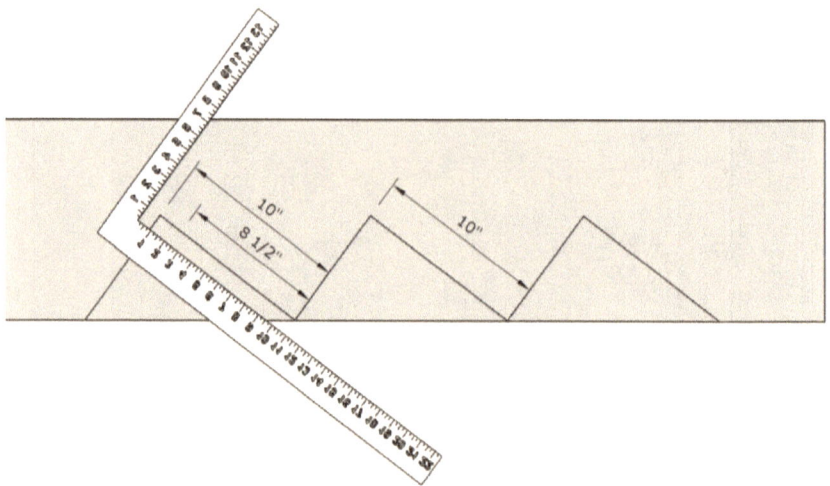

In the next step we will subtract the thickness of the ledger from the top of the stair stringer that will attach to it.

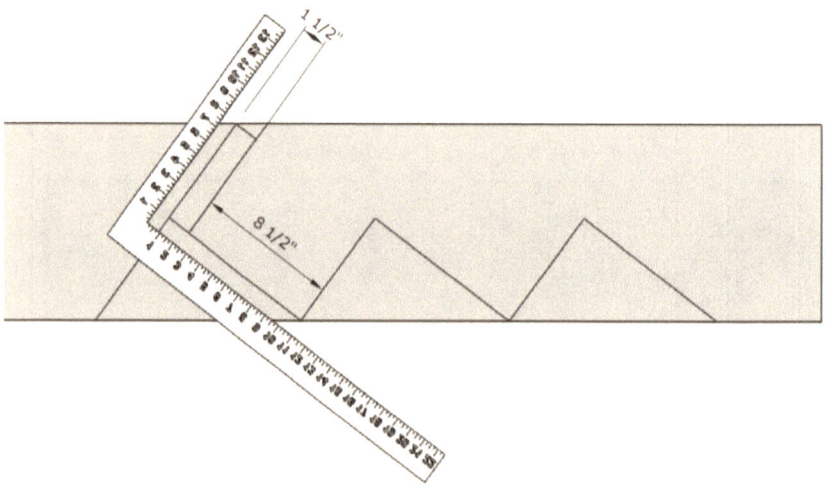

In the illustration above I drew a side view that represents the ledger that's one and a half inch thick to provide you with another example or reason why you need to deduct the width of the ledger.

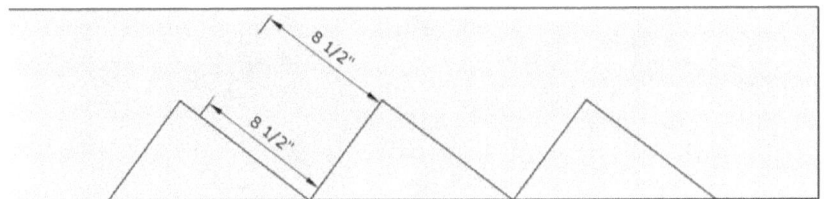

Simply make two marks that parallel the riser as shown in picture above. Remember, if you're using different materials that are wider or thinner than this measurement then it will need to be changed.

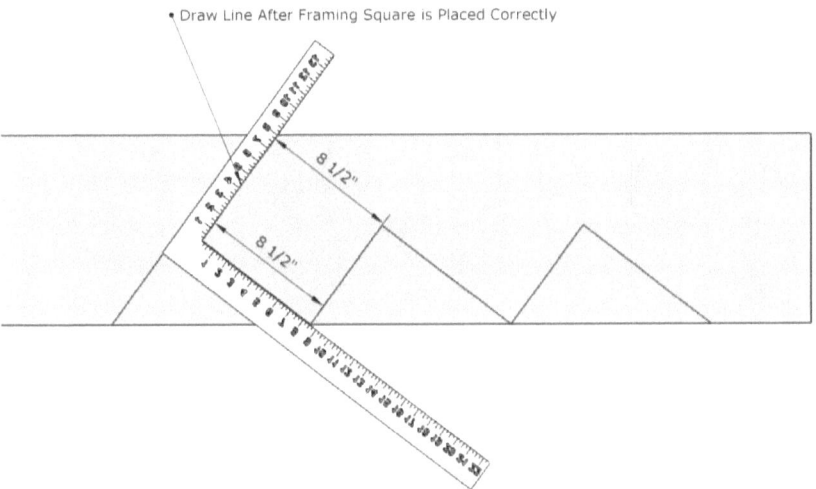

Draw a straight line after framing square is placed correctly and double check the measurement to make sure it's accurate. In other words take your time marking the stair stringer pattern and double check all of your measurements.

It only takes one incorrect measurement to ruin the entire stair stringer layout process.

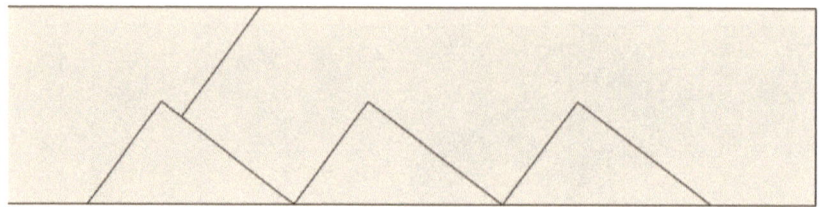

You should end up with something that looks like the illustration above.

Laying Out Bottom

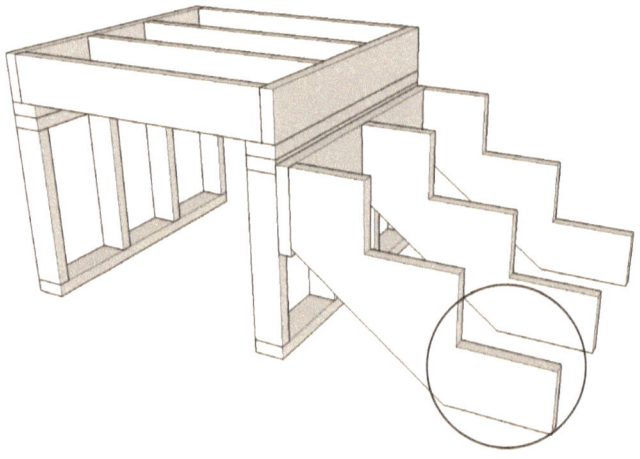

In this section I will provide you with two different methods for laying out the bottom of your stair stringer pattern. One will be for a wood framed floor and the other will be for a concrete slab.

The only difference between the two will be subtracting an additional measurement to allow for a baseplate that will attach to the concrete with framing anchors.

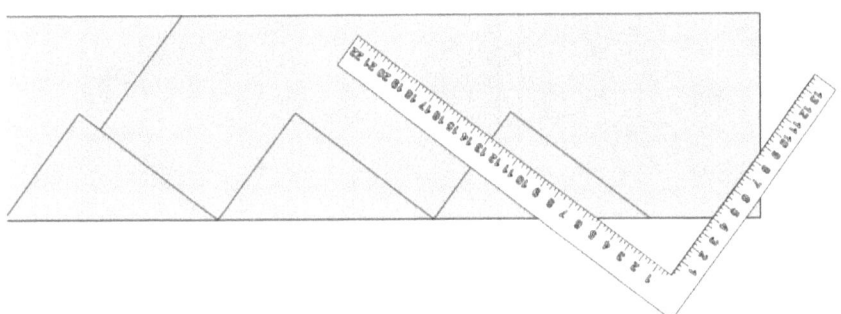

Rotate and place the framing square on top of the stair stringer pattern as shown in illustration above.

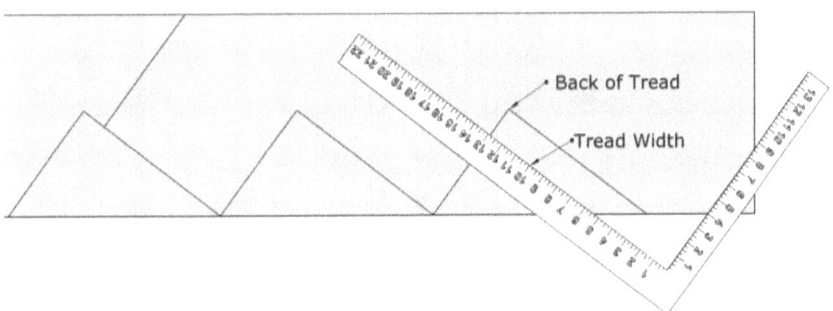

Then line the tread width mark on the framing square up with the back of the tread mark on the pattern and draw a line for your first riser.

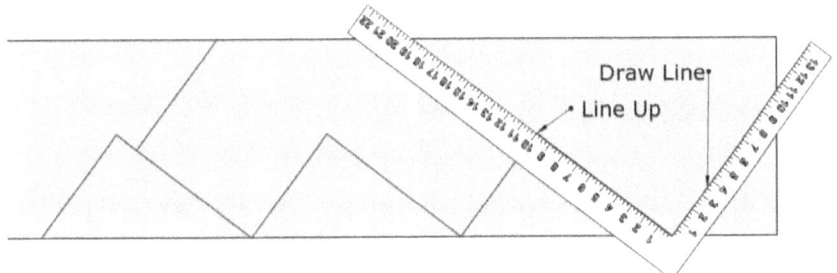

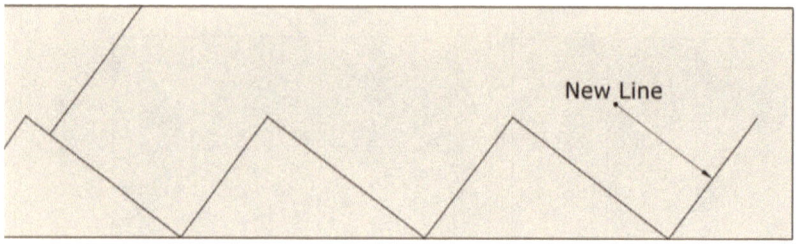

You should end up with something like this.

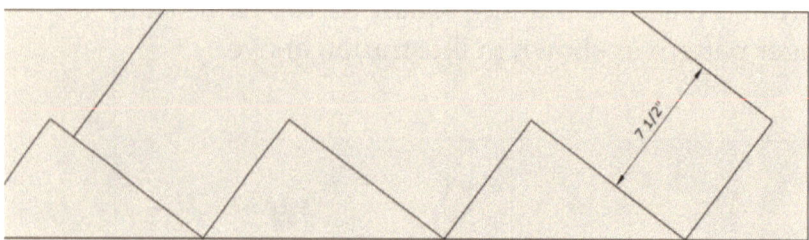

Even though you don't need to draw this line, it might be helpful so go ahead and draw a riser height line parallel to the bottom step anyway. It will be used as a reference point for further clarification.

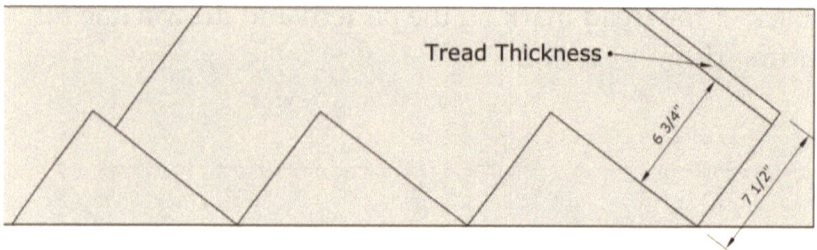

Subtract tread thickness from bottom of stair stringer. In our example we will be using three-quarter inch thick materials for our stair treads.

The previous method can be used if placing bottom of stair stringer directly on top of a wood framed floor, landing or platform, but might not be the best method if sitting directly on top of concrete.

For concrete you can place a piece of 2 x 4 treated in between the top of the concrete floor that will need to be attached to concrete with some type of fasteners and the bottom of the stringer as recommended by most building codes.

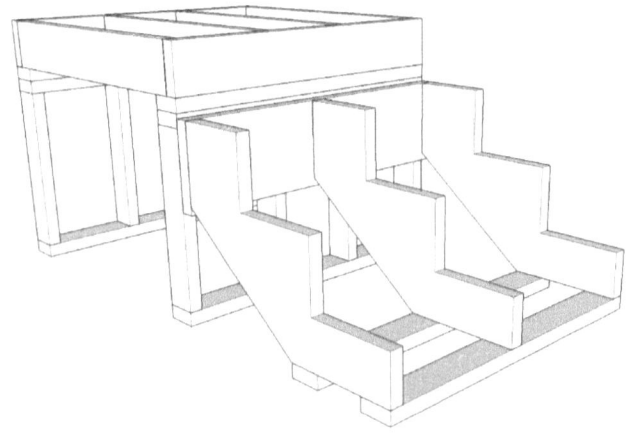

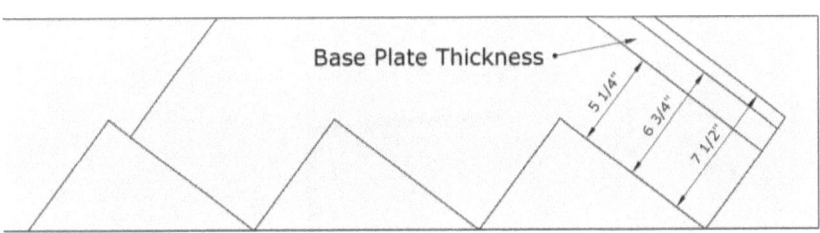

Along with subtracting the thickness of the stair tread, you will also need to subtract the thickness of the materials used for your baseplate.

Starting with a 7 1/2 inch total individual riser height, you will need to subtract 3/4 inches (tread thickness) plus 1 1/2 inches (base plate thickness) from the overall bottom riser height measurement, leaving you with 5 1/4 inches.

After you have cut your stair stringer it wouldn't be a bad idea to place it in position and check it to make sure the steps are level. In other words, don't use it as a pattern to cut the other stair stringers until you make sure it's going to work.

That's it for basic stair stringer layout, but if you're having a difficult time trying to understand why you need to deduct for the stair tread or ledger thickness then the next section might provide you with a better explanation.

Reason for Adjusting Bottom of Stringer

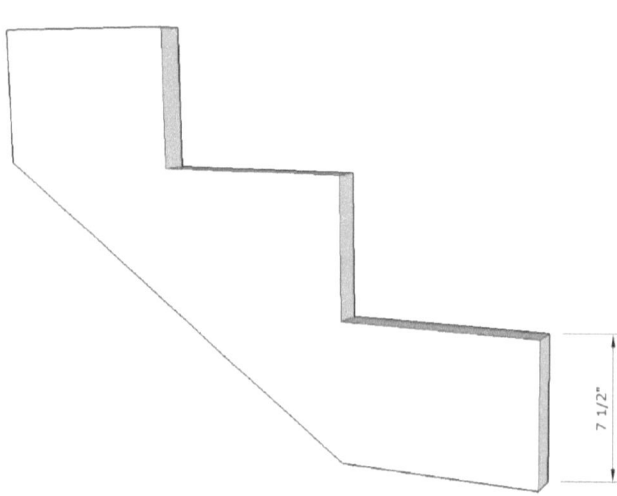

If we don't do anything to the bottom riser and leave it alone, let's see what happens when we add our tread to the top.

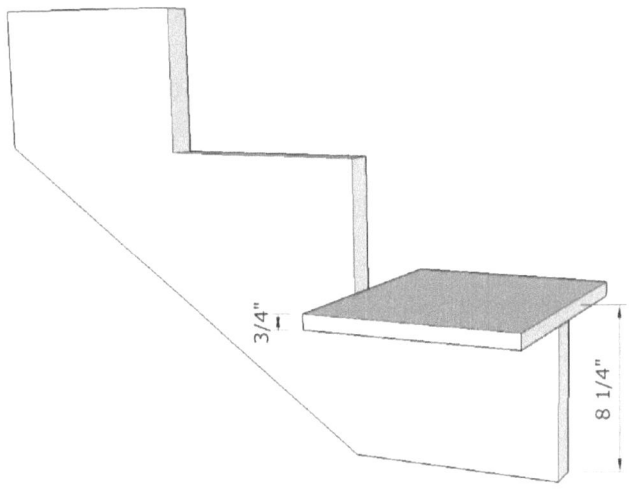

We end up with a taller riser.

This is a common mistake made by do-it-yourselfers and new stair builders, otherwise I wouldn't mention it or provide you with further examples. In order to build stairs correctly, you've got to understand the fundamentals and reasons why you're going to deduct the thickness of the stair tread from the bottom of the stringer.

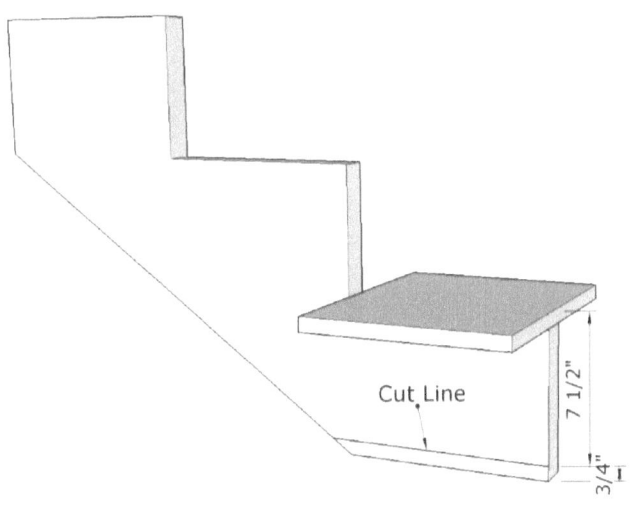

By deducting the thickness of the stair tread from the bottom of the stair stringer, you will be lowering every step along the way. This will put the top of every tread in the correct position.

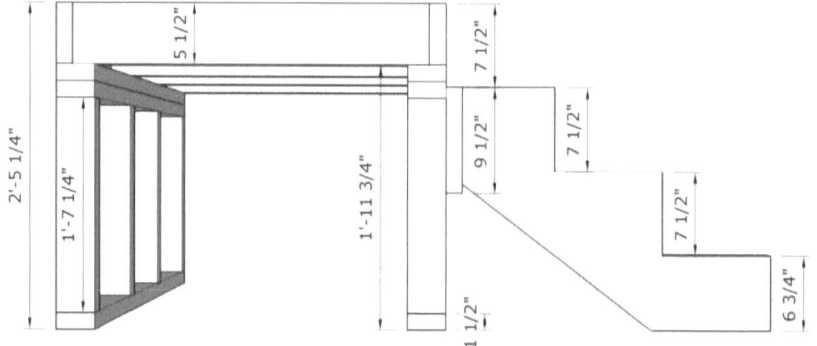

The illustration above provides you with vertical measurements and the illustration below provide you with horizontal measurements for this stairway to give you a better idea how the stairway will be constructed.

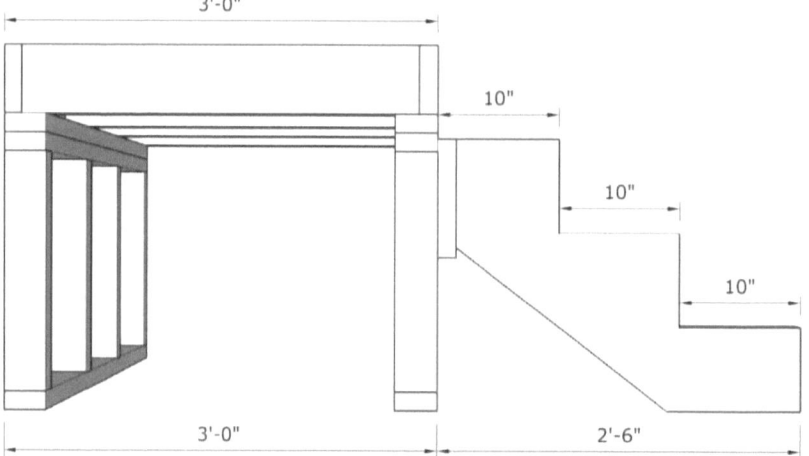

Now pay attention to how these measurements change as we add our treads, risers and landing sheathing.

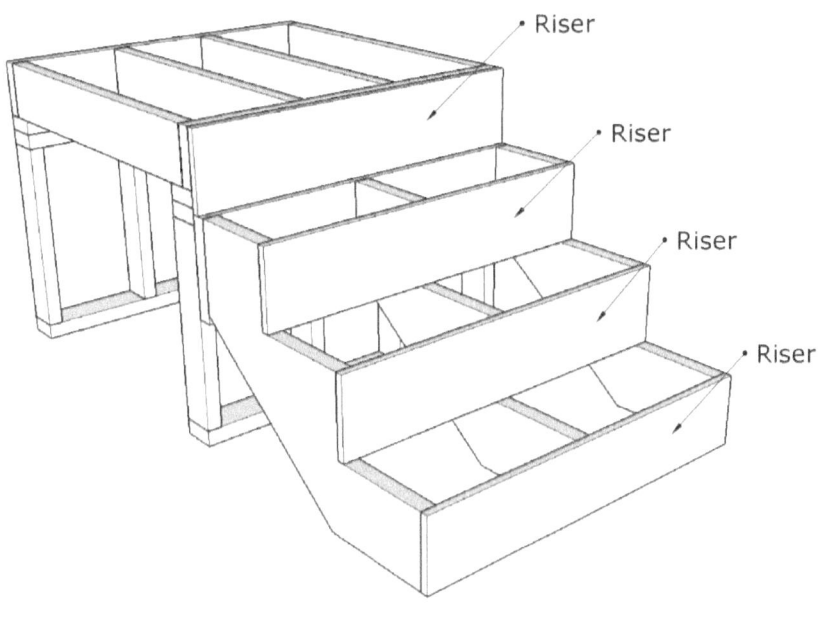

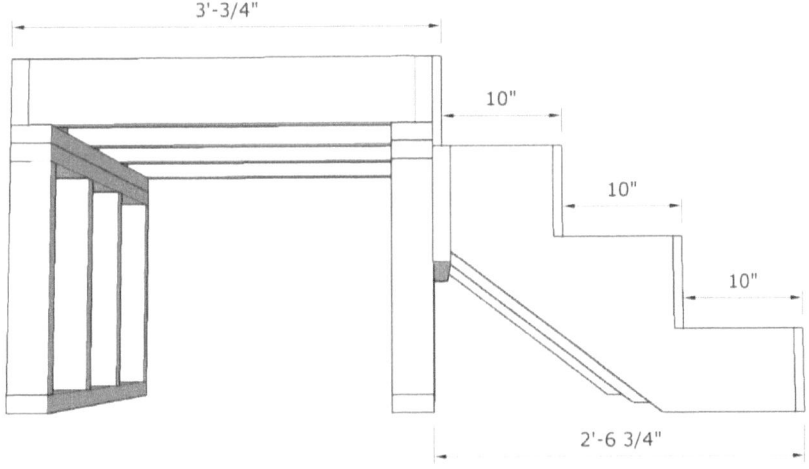

The entire stairway grew by three quarters of an inch horizontally, simply by adding three-quarters of an inch thick risers.

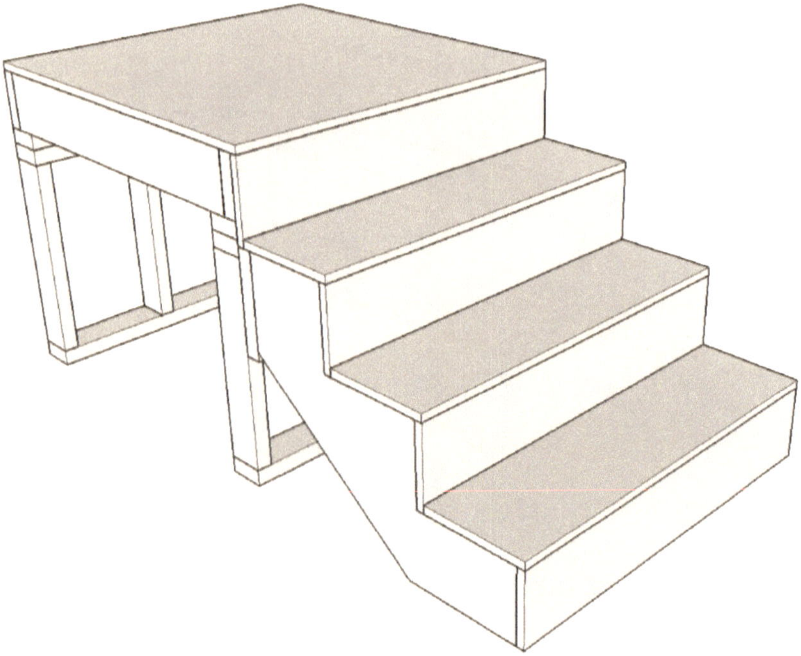

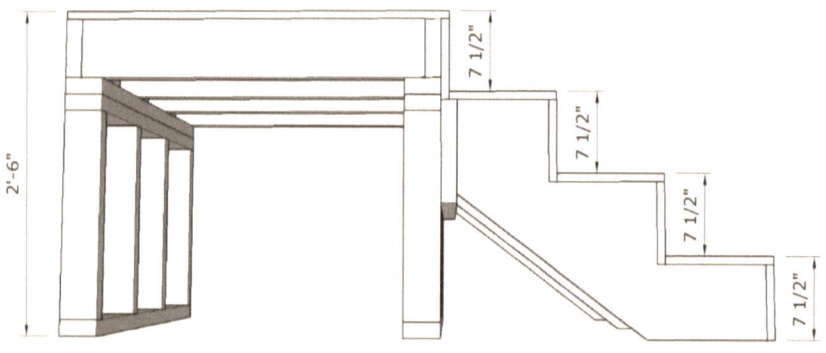

The measurements change again after adding the three-quarter inch thick stair treads and creates equal riser heights. In the next example we are going to add a 1 inch nosing to the front of the stair treads and landing sheathing, changing the horizontal measurements one more time.

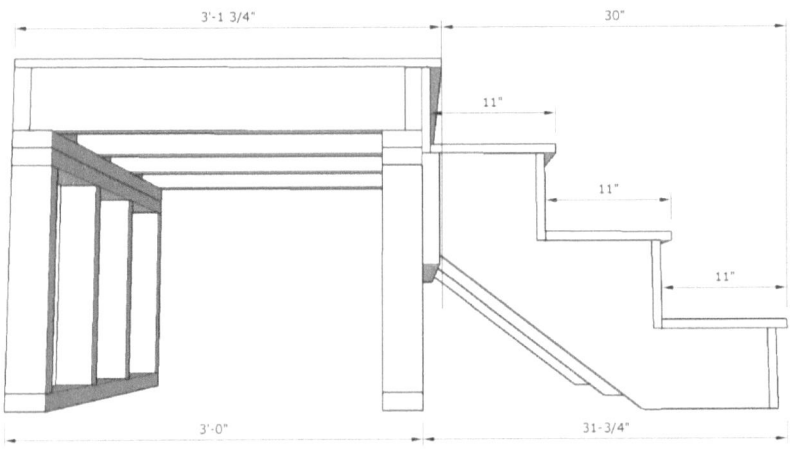

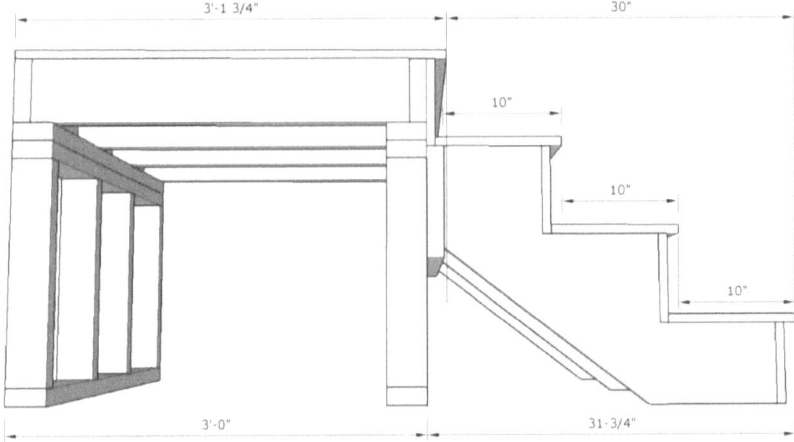

We basically extended the stairway one and three-quarter inches with these measurements and it could be a problem. If you're working in tight areas where every inch is critical, then you will need to move the landing back one and three-quarter inches, so you're finished measurements will work out correctly.

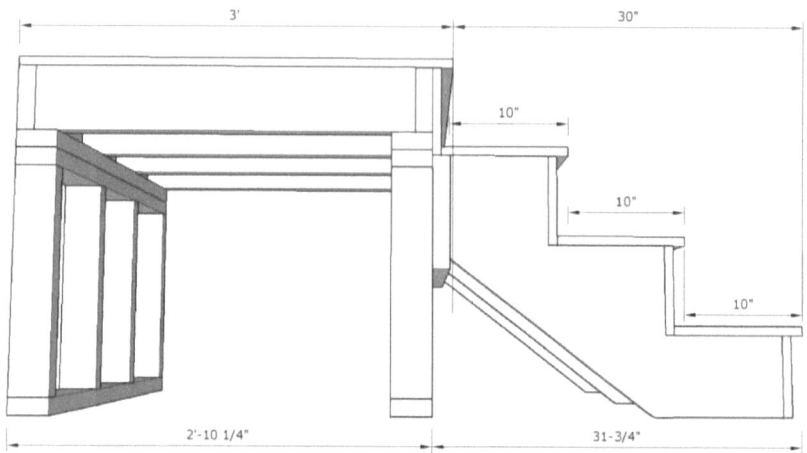

Hopefully by now you have a "Really Good Understanding" of the stair stringer layout process, but if you need a little more information, then watch some of the videos we made by visiting the link below.

http://www.video.stairs4u.com/stair_stringer_layout_videos.html

Notes
_____
_____
_____
_____
_____
_____
_____
_____
_____
_____
_____
_____

Installing Stair Stringers

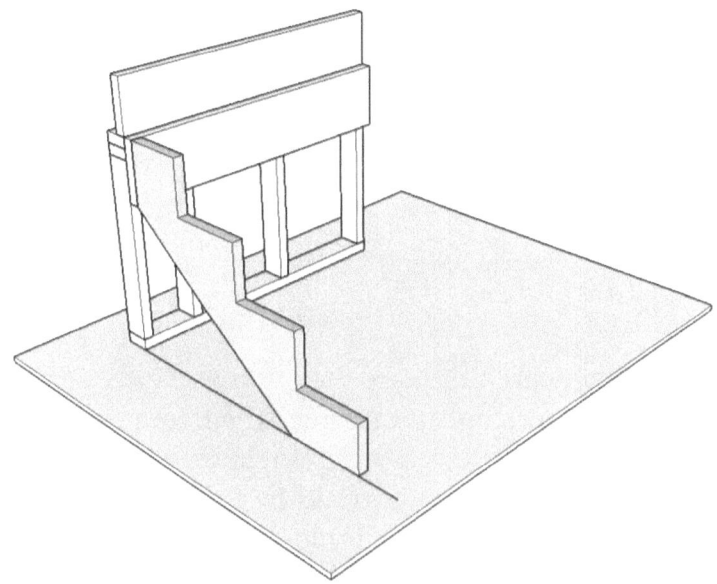

You can start with either the right or left side, just don't start in the center. In the illustration above we're starting with the left side and it can be fastened securely to the top framing members by toe nailing with 16d nails (left below) or using hardware and their recommended fasteners (right below).

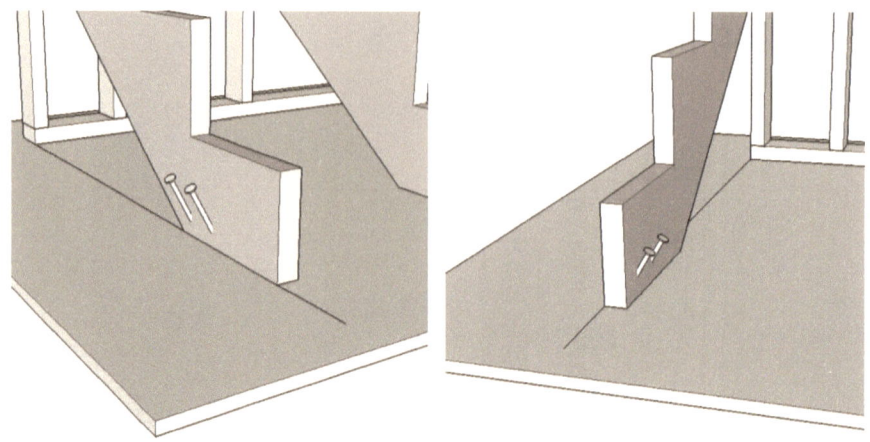

If the stair stringers aren't square then you will have a difficult time locating the correct positioning for the treads and risers. The line on the floor represents a 90° angle from the upper level supporting wall and now would be a good time to make sure you fasten the first stringer as close to 90° as possible.

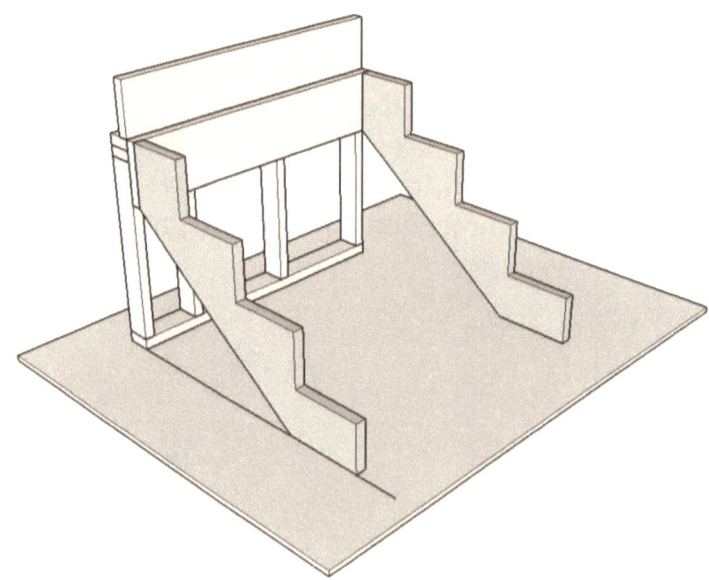

Attach next stringer to other side, just don't nail either one off permanently until you have verified they are parallel to each other.

You can temporarily fasten each stringer with one nail at the top and one nail at the bottom, without standing on them until you have used a straight edge to make sure they are in perfect alignment or as close to it as possible.

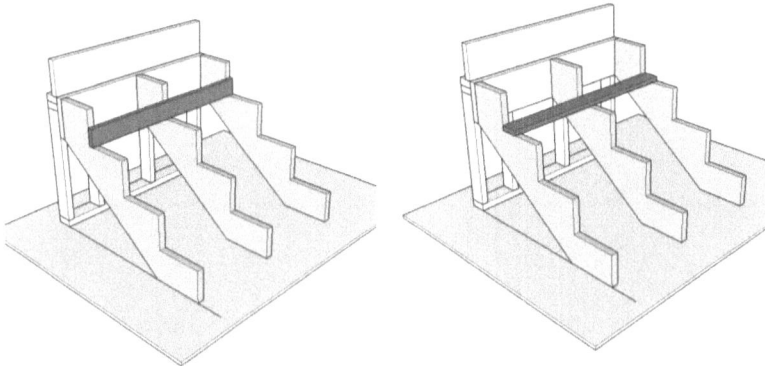

You can use a straight board, riser or level as a straight edge as long as it's long enough to reach from end to end and you can move it into different positions as shown in illustrations above to see if the stringers line up.

You can use shims to adjust the center or side stringers by placing them in between the upper and lower framing members were they connect if they need to be pushed up or forward or remove some of the materials by re-cutting the stringers where they connect to either the lower or upper framing members if the they need to go down or back.

If you do a good job cutting the stair stringers then you shouldn't have a problem lining them up.

For more information visit link below and look for video with the title: Adjust Center Stringers Before Nailing Them Off

http://www.video.stairs4u.com/how_to_assemble_stairway_videos.html

Laying Out Treads and Risers

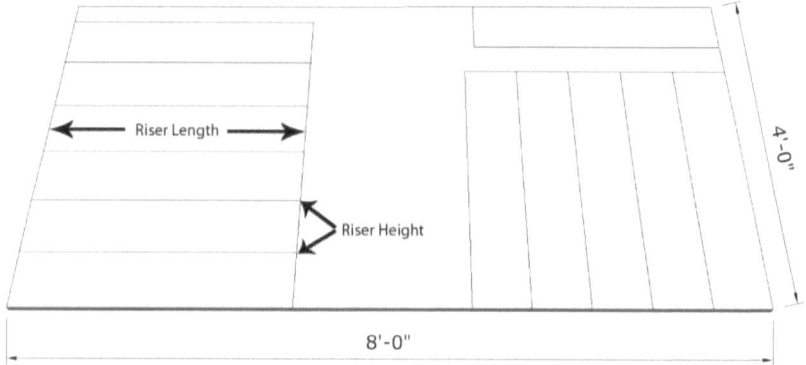

The illustration above provides you with an excellent example of how to layout your risers. You can use a straight edge or chalk line and risers can be laid out either parallel or perpendicular to the 8 foot side of a 4 x 8 sheet of plywood or oriented strand board.

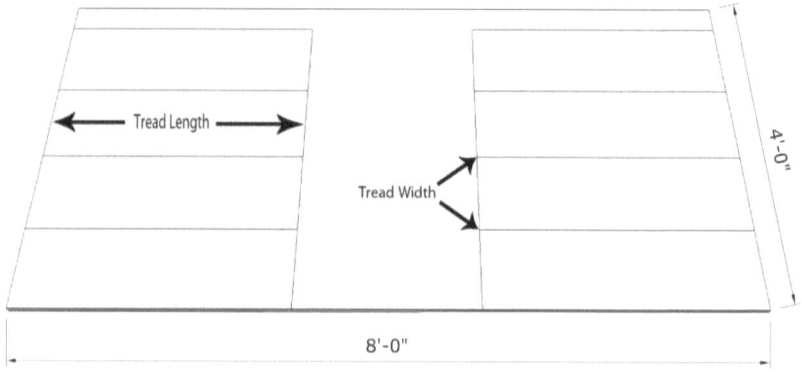

Treads on the other hand should only be laid out parallel to the 8 foot side of a 4 x 8 sheet of plywood. You get more strength this way, but some plywood manufacturers will let you run the treads in either direction as long as you follow their particular installation instructions.

Installing Risers

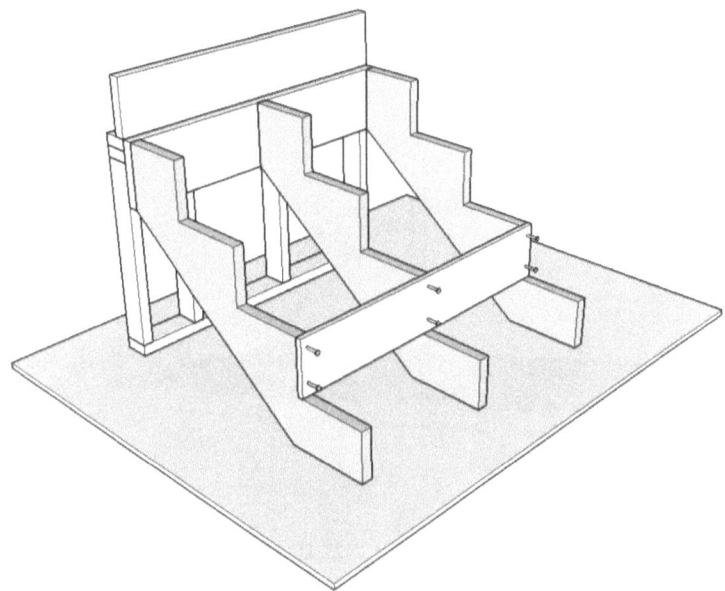

Use two 8d nails for three-quarter inch thick materials and two 16d nails for inch and a half thick materials.

Try to avoid standing on stair stringers even when they are fasten securely. I've stood on them before when large cracks allowed the tips to break and almost fell.

You can nail all of the risers on and then nail all of the treads on or you can play it safe and start from the bottom by fastening the first two lower risers to the stringers, then fastening the first lower tread to the stringers and then work your way up the stairway fastening the next riser, then tread, then riser then tread until you have completed the project.

It's a common construction practice that I'm in full agreement on to nail the risers on first, but there are no rules set in stone that might suggest otherwise. Feel free to nail your treads on first and then the risers if you need to be different.

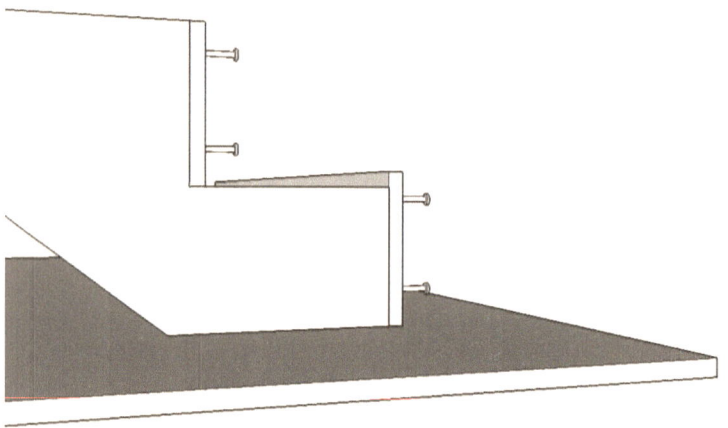

The bottom riser will need to be cut to allow for the tread thickness.

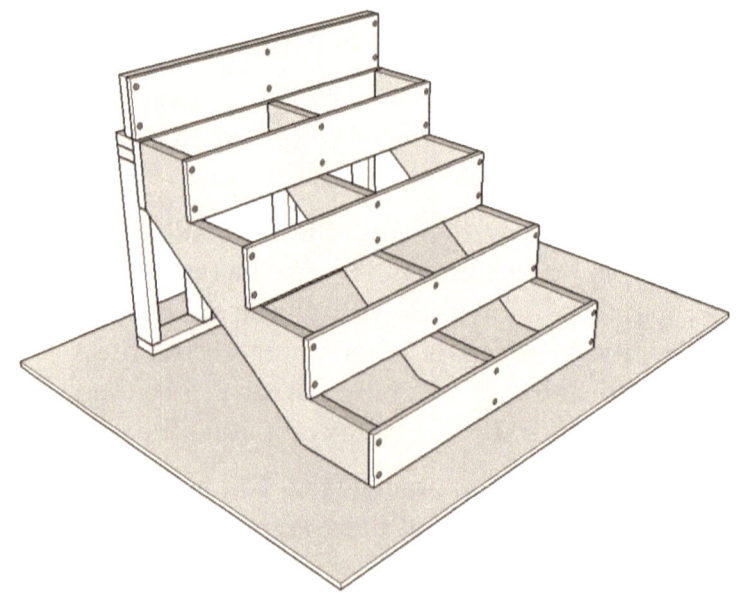

Installing Treads

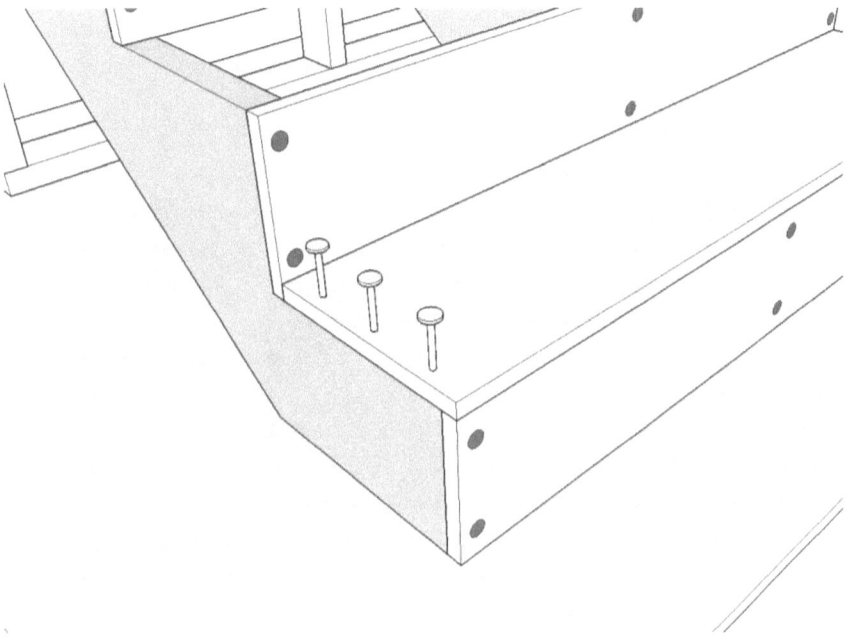

Use 8d nails for three-quarter inch thick materials and 16d nails for inch and a half thick materials. The number of nails used will vary with the width and sizes of the materials. I recommend using three nails for 11 inch wide treads as shown in the illustration above.

When you nail the risers on first and then the treads, the riser will end providing additional structural strength for the front edge of the step.

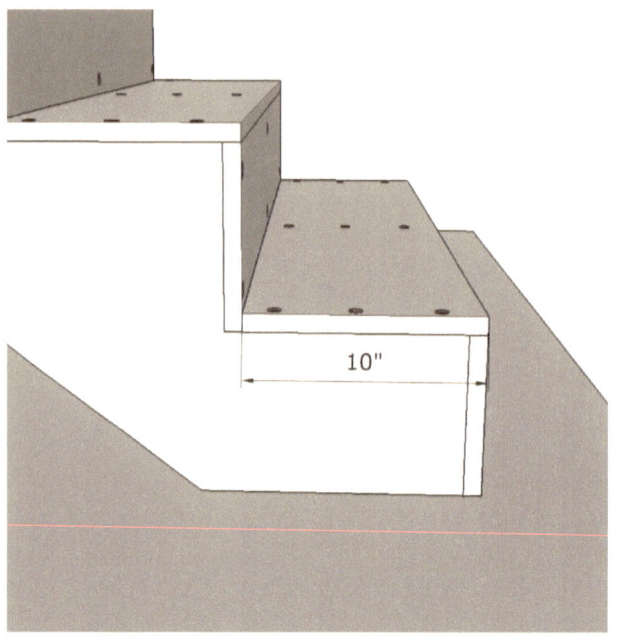

If you want a 1 inch nosing (illustration below) then simply add 1 inch to the width of your stair treads.

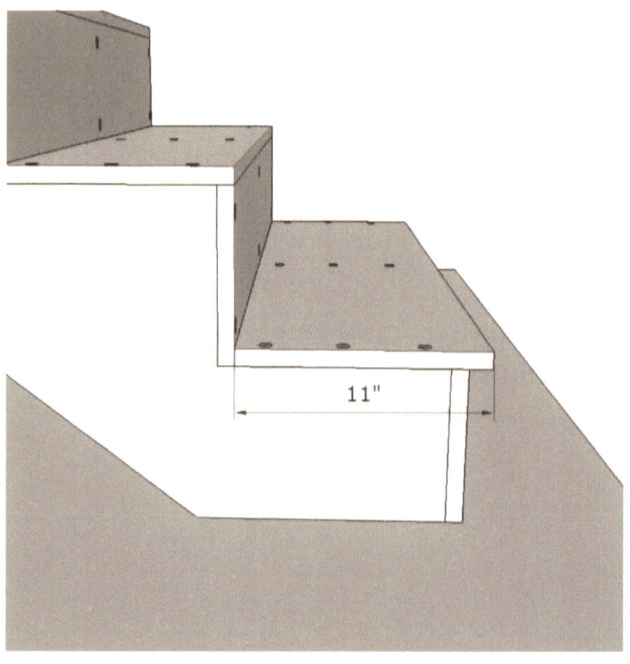

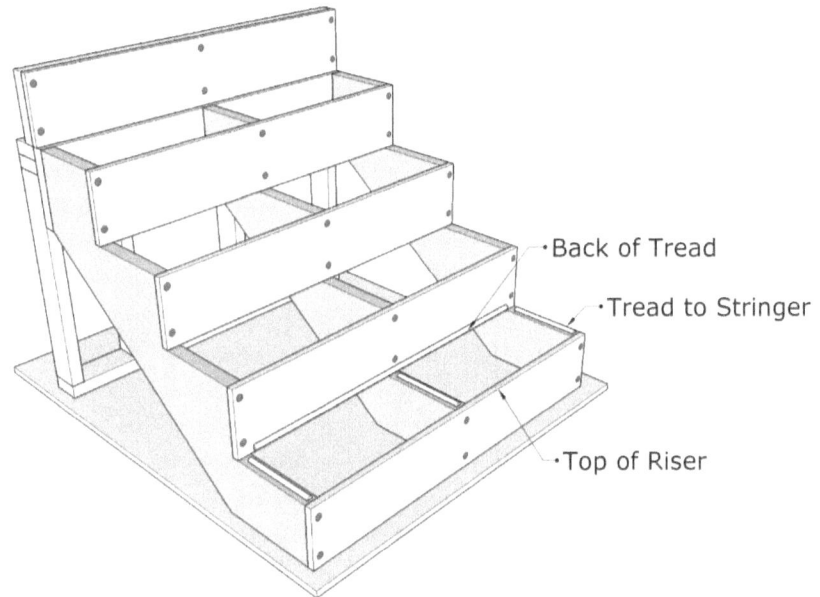

If you're looking for a little more insurance to reduce or even eliminate the chances of your stairs squeaking, then I recommend using adhesive in these areas to provide you with a better connection.

The areas where stairs seem to squeak the most are at the tread to stringer connection point and back of tread. You could always use screws, staples or nails instead of adhesive to fasten the treads to the risers or the risers to the treads, but you should always use adhesive at the tread to stringer connection point.

You can use other materials for treads and risers, like construction standard lumber (2x8, 2x10 or 2x12) that can be easily cut or ripped to length.

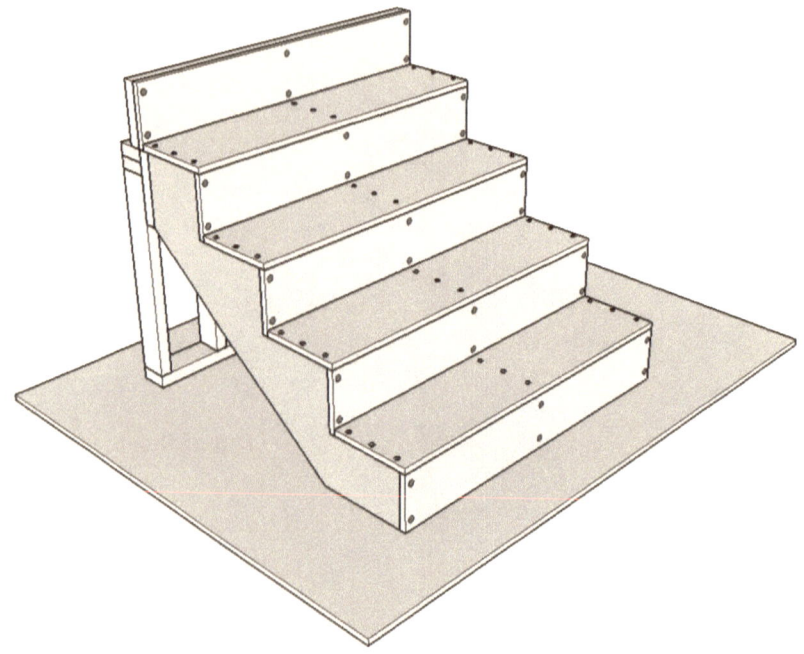

If you end up with something that looks like this then you should pat yourself on the back because you've done something most of the carpenters I worked with for years couldn't do.

Great Job!

Stair Building Glossary

Visit our online glossary for more information or if there's a word we might have missed at

http://stairs4u.com/glossary.htm

Anchor Bolts – Bolts that are embedded in the building foundation and used to attach building materials, like lumber to the foundation. Most anchor bolts are L-shaped and between 1/2 and 3/4" in diameter.

Balusters - These are vertical handrail components used to protect people from falling through the handrail or guard rail system. Balusters can be made from wood, metal and plastic.

Building Codes - A system of rules and regulations to create safer buildings. Most cities or counties throughout the United States regulate building codes through their building departments using building inspectors and other city officials.

Building Foundation - This is usually what a house or building rests upon. Foundations are usually built using concrete, block, cement or brick.

Building Hardware - Any nails, screws, nuts, bolts, hangers, connectors or other metal brackets used to connect one building component to another.

Cap - A piece of finish building material that usually runs at an angle, vertical or horizontal. Most solid drywall guardrails will use a 1 x 6 cap to finish the top of a wall off.

Cedar – Is a recommended wood used for exterior building projects like decks and stairs. Cedar can also be used on the interior of a house. Also see Redwood.

Ceiling - The overhead surface of a room, hallway and stairway. It's important to have enough distance between the ceiling and your head, for anyone who'll be walking up and down the stairs.

Construction Standard Lumber - Wood that is used primarily for framing and building homes. On the west coast of the United States they use Douglas Fir and on the East Coast they often use Southern Pine.

Diameter - The width of a circle, using a straight line that runs directly through the center.

Douglas Fir – A tall straight growing tree that provides plenty of lumber used for building homes throughout most western areas of the United States. Most construction standard lumber used in Southern California is Douglas Fir.

Drywall Spacer - Usually a 2 x 4 or 1 x 4 that attaches to one side of the stair stringer to allow a gap between the wall and the stringer to allow easy drywall installation around the stair stringer's.

First Floor - This would be the lower level of the building. The first floor would be a lower living area of a two-story house. The bottom of the stairway usually attaches to the first floor.

Framing Anchors - These can be made from lead, plastic or metal and are used to attach one type of building material to another. A good example of this would be using mechanical anchoring systems to attach treated lumber to a concrete building foundation.

Gripable Handrailing - This is the part of the handrailing system that's used to hold on to, while you're walking up and down a stairway. Most gripable handrails are located between 34 and 38 inches from the top of the stair tread nosing and run the entire length of the stairway.

Guard Railing - Consists of balusters, handrailing and posts used to create a protective barrier to prevent people from falling from heights exceeding 30 inches, above the lower level.

Head Room - This is the vertical distance between the stair tread nosing and the ceiling above. This building code prevents taller people from hitting their heads on the ceiling, as they walk up and down the stairway.

Headout - Is located in the upper level floor joist system, where the stairway attaches to the upper floor. The stair stringers usually attach to the headout or a ledger, nailed directly underneath the headout.

Jack Studs - These are vertical wall studs that form the walls underneath a stair stringer. The bottom cut of the jack studs are usually square, while the top cut is angled.

Joist - These are the horizontal structural support members of the floor framing system that are used to support the floor. Floor joist are usually 2 x 6, 2 x 8, 2 x 10, 2 x 12 or 2 x 14 and also come in the form of engineered lumber, like truss joist.

Joist Hanger - These are the metal connectors used to connect the floor joist to another structural component of the floor joist system, like a beam. You can also use joist hangers to connect the stair stringers to a ledger or headout.

Lag Screws - Are screw type bolts with pointed tips that are used to attach building components together. Stair builders often use lag screws to attach 3 x 12 treads to 4 x 12 stringers, using special brackets.

Landing - A floor located somewhere in a stairway. A stairway could go up five steps and then turn 90° at a stair landing and then continue up to the second floor. A stair landing could also be located at the bottom or top of the stairway.

Layout - Marking, measuring and planning out the design of walls, floors, roofs or stairway components. Laying out a stair stringer is often referred to as the process for planning and creating the stair stringer, before cutting it.

Ledger - This is usually a horizontal building component that's used to attach two sections of the building together. A good example of this would be using a ledger that attaches to a wall, providing something solid for your stair stringers to attach to.

Nails – Pin shaped pieces of metal that comes in different sizes and can be used to attach wood together. Most stairway framers use 16-d and 8-d nails for stair building.

Newel Post - Usually decorative wood posts used for supporting the handrailing system, located at the top, middle or bottom of a stairway. Newel posts often provide structural support for stairway handrails.

Nosing - This is the front of the stair tread or step. Some stair steps have a nosing that protrudes 1 inch away from the riser, while others don't protrude out at all, yet angle back in.

O.S.B. - Is often referred to as oriented strand board. This is a man-made product that requires gluing strands of lumber together, to produce an alternative building product to plywood.

Pattern - An original design created for the purpose of making copies. You can use patterns to make copies of stair stringers, treads and risers to speed up the stair assembly process.

Plywood - Is a man-made product made by gluing thin layers of lumber together with each layer changing direction at 90° angles, increasing the strength of the product as additional layers are glued to it. Plywood is often used for floors, structural walls, stair treads and risers.

Posts - Structural components that provide support for a beam or strength for handrails. Guard rails usually have posts built into them, to provide additional strength.

Radius - Is half the diameter of a circle. The radius is also the measurement from the center to the outside of a circle.

Railing - This is normally referred to as the top of the handrail or guard rail system. If you were walking next to a guard rail, you could run your fingers along the top of the railing.

Redwood - Is a preferred wood used for building outdoor decks and stairways. Redwood comes in different grades and some of these grades are extremely expensive. I've heard other people say that termites won't eat Redwood, but that's not true.

Risers - These are the vertical sections of the stairway in between the treads or steps. Risers can be made from a variety of materials, but the most common are made from construction standard, hardwoods, soft woods, plywood's and oriented strand board.

Screw - These are pieces of pin shaped metal with a threaded body that are often used to assemble home building products. Screws are often used instead of nails to provide additional holding power. Don't replace screws with nails, unless you have permission from the building designer, architect or engineer. Screws usually don't have the same sheer value a nail does.

Second Floor - This is usually the second level of a building. The top of the first stairway will attach to the second floor, while the bottom of the stairway attaches to the first level or floor. If you have a three level house, the bottom of the second set of stairs will attach to the second floor, while the top of the second stairway attaches to the third floor.

Sheathing - This is the material you will stand on, while walking around on a wood framed floor system. Sheathing is often referred to as underlayment. Most home builders use plywood or oriented strand board for sheathing.

Skirt Board - There are two types of skirt boards. The first type goes on the outside of the stair stringer and the second type separates your finished wall from the stairway. Both types of skirt boards are finished building products and are normally stained or painted.

Span - This would be the distance in between structural supports. The span will determine the thickness of stair treads, risers and floor sheathing.

Stairway - A passageway used to gain access from one level of the building to another. To get from the first floor of most houses to the second floor, a stairway is provided by the home builder for easy access.

Stairwell - This is an interior shape that's usually cut into a buildings floor framing system, to provide access to the second floor. The framing carpenters who are responsible for building the second floor are also responsible for building the stairwell.

Stringer - The main structural support for almost any stairway. Stair stringers can be made out of wood or metal and usually support each tread and riser (step).

Stringer Layout – This is the process for marking out each individual stair tread and riser, on the stair stringer. This usually requires a framing square, along with the height of each riser and the width of each stair tread.

Total Rise - This is the vertical distance from the top of the bottom floor to the top of the top floor. If you had a 100 inch measurement from the top of the building foundation (lower level), to the top of the second floor sheathing, then the total rise would be 100 inches.

Total Run - This is the horizontal distance from the front of the first stair step, to the back of the last stair step. If you went horizontally from the stair headout, to the first floor and measured the distance to the first step, this would be the total run.

Tread Brackets - These are brackets used to attach stair treads to stair stringers. Most of these tread brackets can be purchased from your local home improvement center or lumber yard.

Tread Overhang - This is the distance each stair tread overhangs from the riser. Most building codes won't let you have a tread overhang longer than 1 - 1/4 inches.

Tread Under Cut - This is the distance the stair tread protrudes into the stair riser. Instead of an overhang at the front of the stair step, the riser will be angled back, at the bottom, providing more room for the stair step or tread.

Tread – Another word for the stair step, often used by building professionals. Stair treads can be made from a variety of materials, including concrete, wood and metal.

Treated Lumber - Is wood that has been pressure-treated with chemicals to reduce the chances of wood decay. Treated lumber is often used at the bottom of wood framed walls and stair stringers, to reduce the chances of wood rot or decay.

Visit our online glossary for more information or if there's a word we might have missed at:

http://stairs4u.com/glossary.htm

Rise And Run Chart For 10 Inch Treads

Total Rise = Total height of stairway, from the top of the bottom floor, to the top of the top floor.

Risers = Amount of risers in stairway.

Riser Height = Individual riser height or total height between steps.

Steps = Amount of steps in stairway.

Total Run = Length of stairway in inches, using a 10 inch step or stair tread.

Angle = This is the angle of the stairway or incline.

Total Rise	Risers	Riser Height	Steps	Total Run	Angle
10	2	5.00	1	10	26.57
10.25	2	5.13	1	10	27.14
10.5	2	5.25	1	10	27.70
10.75	2	5.38	1	10	28.26
10.75	2	5.38	1	10	28.26
11.25	2	5.63	1	10	29.36
11.5	2	5.75	1	10	29.90
11.75	2	5.88	1	10	30.43
12	2	6.00	1	10	30.96
12.25	2	6.13	1	10	31.49
12.5	2	6.25	1	10	32.01
12.75	2	6.38	1	10	32.52
13	2	6.50	1	10	33.02
13.25	2	6.63	1	10	33.52
13.5	2	6.75	1	10	34.02
13.75	2	6.88	1	10	34.51
14	2	7.00	1	10	34.99
14.25	2	7.13	1	10	35.47
14.5	2	7.25	1	10	35.94
14.75	2	7.38	1	10	36.41

Total Rise	Risers	Riser Height	Steps	Total Run	Angle
15	2	7.50	1	10	36.87
15.25	2	7.63	1	10	37.33
15.5	2	7.75	1	10	37.78
15.75	3	5.25	2	20	27.70
16	3	5.33	2	20	28.07
16.25	3	5.42	2	20	28.44
16.5	3	5.50	2	20	28.81
16.75	3	5.58	2	20	29.18
17	3	5.67	2	20	29.54
17.25	3	5.75	2	20	29.90
17.5	3	5.83	2	20	30.26
17.75	3	5.92	2	20	30.61
18	3	6.00	2	20	30.96
18.25	3	6.08	2	20	31.31
18.5	3	6.17	2	20	31.66
18.75	3	6.25	2	20	32.01
19	3	6.33	2	20	32.35
19.25	3	6.42	2	20	32.69
19.5	3	6.50	2	20	33.02
19.75	3	6.58	2	20	33.36
20	3	6.67	2	20	33.69
20.25	3	6.75	2	20	34.02
20.5	3	6.83	2	20	34.35
20.75	3	6.92	2	20	34.67
21	3	7.00	2	20	34.99
21.25	3	7.08	2	20	35.31
21.5	3	7.17	2	20	35.63
21.75	3	7.25	2	20	35.94
22	3	7.33	2	20	36.25
22.25	3	7.42	2	20	36.56
22.5	3	7.50	2	20	36.87
22.75	3	7.58	2	20	37.17
23	3	7.67	2	20	37.48
23.25	3	7.75	2	20	37.78
23.5	4	5.88	3	30	30.43
23.75	4	5.94	3	30	30.70
24	4	6.00	3	30	30.96
24.25	4	6.06	3	30	31.23
24.5	4	6.13	3	30	31.49
24.75	4	6.19	3	30	31.75
25	4	6.25	3	30	32.01

Total Rise	Risers	Riser Height	Steps	Total Run	Angle
25.25	4	6.31	3	30	32.26
25.5	4	6.38	3	30	32.52
25.75	4	6.44	3	30	32.77
26	4	6.50	3	30	33.02
26.25	4	6.56	3	30	33.27
26.5	4	6.63	3	30	33.52
26.75	4	6.69	3	30	33.77
27	4	6.75	3	30	34.02
27.25	4	6.81	3	30	34.26
27.5	4	6.88	3	30	34.51
27.75	4	6.94	3	30	34.75
28	4	7.00	3	30	34.99
28.25	4	7.06	3	30	35.23
28.5	4	7.13	3	30	35.47
28.75	4	7.19	3	30	35.71
29	4	7.25	3	30	35.94
29.25	4	7.31	3	30	36.18
29.5	4	7.38	3	30	36.41
29.75	4	7.44	3	30	36.64
30	4	7.50	3	30	36.87
30.25	4	7.56	3	30	37.10
30.5	4	7.63	3	30	37.33
30.75	4	7.69	3	30	37.55
31	4	7.75	3	30	37.78
31.25	5	6.25	4	40	32.01
31.5	5	6.30	4	40	32.21
31.75	5	6.35	4	40	32.42
32	5	6.40	4	40	32.62
32.25	5	6.45	4	40	32.82
32.5	5	6.50	4	40	33.02
32.75	5	6.55	4	40	33.22
33	5	6.60	4	40	33.42
33.25	5	6.65	4	40	33.62
33.5	5	6.70	4	40	33.82
33.75	5	6.75	4	40	34.02
34	5	6.80	4	40	34.22
34.25	5	6.85	4	40	34.41
34.5	5	6.90	4	40	34.61
34.75	5	6.95	4	40	34.80
35	5	7.00	4	40	34.99
35.25	5	7.05	4	40	35.18

Total Rise	Risers	Riser Height	Steps	Total Run	Angle
35.5	5	7.10	4	40	35.37
35.75	5	7.15	4	40	35.56
36	5	7.20	4	40	35.75
36.25	5	7.25	4	40	35.94
36.5	5	7.30	4	40	36.13
36.75	5	7.35	4	40	36.32
37	5	7.40	4	40	36.50
37.25	5	7.45	4	40	36.69
37.5	5	7.50	4	40	36.87
37.75	5	7.55	4	40	37.05
38	5	7.60	4	40	37.23
38.25	5	7.65	4	40	37.42
38.5	5	7.70	4	40	37.60
38.75	5	7.75	4	40	37.78
39	6	6.50	5	50	33.02
39.25	6	6.54	5	50	33.19
39.5	6	6.58	5	50	33.36
39.75	6	6.63	5	50	33.52
40	6	6.67	5	50	33.69
40.25	6	6.71	5	50	33.86
40.5	6	6.75	5	50	34.02
40.75	6	6.79	5	50	34.18
41	6	6.83	5	50	34.35
41.25	6	6.88	5	50	34.51
41.5	6	6.92	5	50	34.67
41.75	6	6.96	5	50	34.83
42	6	7.00	5	50	34.99
42.25	6	7.04	5	50	35.15
42.5	6	7.08	5	50	35.31
42.75	6	7.13	5	50	35.47
43	6	7.17	5	50	35.63
43.25	6	7.21	5	50	35.79
43.5	6	7.25	5	50	35.94
43.75	6	7.29	5	50	36.10
44	6	7.33	5	50	36.25
44.25	6	7.38	5	50	36.41
44.5	6	7.42	5	50	36.56
44.75	6	7.46	5	50	36.72
45	6	7.50	5	50	36.87
45.25	6	7.54	5	50	37.02
45.5	6	7.58	5	50	37.17

Total Rise	Risers	Riser Height	Steps	Total Run	Angle
45.75	6	7.63	5	50	37.33
46	6	7.67	5	50	37.48
46.25	6	7.71	5	50	37.63
46.5	6	7.75	5	50	37.78
46.75	7	6.68	6	60	33.74
47	7	6.71	6	60	33.88
47.25	7	6.75	6	60	34.02
47.5	7	6.79	6	60	34.16
47.75	7	6.82	6	60	34.30
48	7	6.86	6	60	34.44
48.25	7	6.89	6	60	34.58
48.5	7	6.93	6	60	34.72
48.75	7	6.96	6	60	34.85
49	7	7.00	6	60	34.99
49.25	7	7.04	6	60	35.13
49.5	7	7.07	6	60	35.27
49.75	7	7.11	6	60	35.40
50	7	7.14	6	60	35.54
5.25	7	0.75	6	60	4.29
50.5	7	7.21	6	60	35.81
50.75	7	7.25	6	60	35.94
51	7	7.29	6	60	36.08
51.25	7	7.32	6	60	36.21
51.5	7	7.36	6	60	36.34
51.75	7	7.39	6	60	36.47
52	7	7.43	6	60	36.61
52.25	7	7.46	6	60	36.74
52.5	7	7.50	6	60	36.87
52.75	7	7.54	6	60	37.00
53	7	7.57	6	60	37.13
53.25	7	7.61	6	60	37.26
53.5	7	7.64	6	60	37.39
53.75	7	7.68	6	60	37.52
54	7	7.71	6	60	37.65
54.25	7	7.75	6	60	37.78
54.5	8	6.81	7	70	34.26
54.75	8	6.84	7	70	34.39
55	8	6.88	7	70	34.51
55.25	8	6.91	7	70	34.63
55.5	8	6.94	7	70	34.75
55.75	8	6.97	7	70	34.87

Total Rise	Risers	Riser Height	Steps	Total Run	Angle
56	8	7.00	7	70	34.99
56.25	8	7.03	7	70	35.11
56.5	8	7.06	7	70	35.23
56.75	8	7.09	7	70	35.35
57	8	7.13	7	70	35.47
57.25	8	7.16	7	70	35.59
57.5	8	7.19	7	70	35.71
57.75	8	7.22	7	70	35.82
58	8	7.25	7	70	35.94
58.25	8	7.28	7	70	36.06
58.5	8	7.31	7	70	36.18
58.75	8	7.34	7	70	36.29
59	8	7.38	7	70	36.41
59.25	8	7.41	7	70	36.52
59.5	8	7.44	7	70	36.64
59.75	8	7.47	7	70	36.76
60	8	7.50	7	70	36.87
60.25	8	7.53	7	70	36.98
60.5	8	7.56	7	70	37.10
60.75	8	7.59	7	70	37.21
61	8	7.63	7	70	37.33
61.25	8	7.66	7	70	37.44
61.5	8	7.69	7	70	37.55
61.75	8	7.72	7	70	37.66
62	8	7.75	7	70	37.78
62.25	9	6.92	8	80	34.67
62.5	9	6.94	8	80	34.78
62.75	9	6.97	8	80	34.89
63	9	7.00	8	80	34.99
63.25	9	7.03	8	80	35.10
63.5	9	7.06	8	80	35.21
63.75	9	7.08	8	80	35.31
64	9	7.11	8	80	35.42
64.25	9	7.14	8	80	35.52
64.5	9	7.17	8	80	35.63
64.75	9	7.19	8	80	35.73
65	9	7.22	8	80	35.84
65.25	9	7.25	8	80	35.94
65.5	9	7.28	8	80	36.05
65.75	9	7.31	8	80	36.15
66	9	7.33	8	80	36.25

Total Rise	Risers	Riser Height	Steps	Total Run	Angle
66.25	9	7.36	8	80	36.36
66.5	9	7.39	8	80	36.46
66.75	9	7.42	8	80	36.56
67	9	7.44	8	80	36.67
67.25	9	7.47	8	80	36.77
67.5	9	7.50	8	80	36.87
67.75	9	7.53	8	80	36.97
68	9	7.56	8	80	37.07
68.25	9	7.58	8	80	37.17
68.5	9	7.61	8	80	37.28
68.75	9	7.64	8	80	37.38
69	10	6.90	9	90	34.61
69.25	10	6.93	9	90	34.70
69.5	10	6.95	9	90	34.80
69.75	10	6.98	9	90	34.90
70	10	7.00	9	90	34.99
70.25	10	7.03	9	90	35.09
70.5	10	7.05	9	90	35.18
70.75	10	7.08	9	90	35.28
71	10	7.10	9	90	35.37
71.25	10	7.13	9	90	35.47
71.5	10	7.15	9	90	35.56
71.75	10	7.18	9	90	35.66
72	10	7.20	9	90	35.75
72.25	10	7.23	9	90	35.85
72.5	10	7.25	9	90	35.94
72.75	10	7.28	9	90	36.04
73	10	7.30	9	90	36.13
73.25	10	7.33	9	90	36.22
73.5	10	7.35	9	90	36.32
73.75	10	7.38	9	90	36.41
74	10	7.40	9	90	36.50
74.25	10	7.43	9	90	36.59
74.5	10	7.45	9	90	36.69
74.75	10	7.48	9	90	36.78
75	10	7.50	9	90	36.87
75.25	10	7.53	9	90	36.96
75.5	10	7.55	9	90	37.05
75.75	10	7.58	9	90	37.14
76	10	7.60	9	90	37.23
76.25	10	7.63	9	90	37.33

Total Rise	Risers	Riser Height	Steps	Total Run	Angle
76.5	11	6.95	10	100	34.82
76.75	11	6.98	10	100	34.90
77	11	7.00	10	100	34.99
77.25	11	7.02	10	100	35.08
77.5	11	7.05	10	100	35.17
77.75	11	7.07	10	100	35.25
78	11	7.09	10	100	35.34
78.25	11	7.11	10	100	35.43
78.5	11	7.14	10	100	35.51
78.75	11	7.16	10	100	35.60
79	11	7.18	10	100	35.69
79.25	11	7.20	10	100	35.77
79.5	11	7.23	10	100	35.86
79.75	11	7.25	10	100	35.94
80	11	7.27	10	100	36.03
80.25	11	7.30	10	100	36.11
80.5	11	7.32	10	100	36.20
80.75	11	7.34	10	100	36.28
81	11	7.36	10	100	36.37
81.25	11	7.39	10	100	36.45
81.5	11	7.41	10	100	36.54
81.75	11	7.43	10	100	36.62
82	11	7.45	10	100	36.70
82.25	11	7.48	10	100	36.79
82.5	11	7.50	10	100	36.87
82.75	11	7.52	10	100	36.95
83	11	7.55	10	100	37.04
83.25	11	7.57	10	100	37.12
83.5	11	7.59	10	100	37.20
83.75	11	7.61	10	100	37.28
84	11	7.64	10	100	37.37
84.25	12	7.02	11	110	35.07
84.5	12	7.04	11	110	35.15
84.75	12	7.06	11	110	35.23
85	12	7.08	11	110	35.31
85.25	12	7.10	11	110	35.39
85.5	12	7.13	11	110	35.47
85.75	12	7.15	11	110	35.55
86	12	7.17	11	110	35.63
86.25	12	7.19	11	110	35.71
86.5	12	7.21	11	110	35.79

Total Rise	Risers	Riser Height	Steps	Total Run	Angle
86.75	12	7.23	11	110	35.86
87	12	7.25	11	110	35.94
87.25	12	7.27	11	110	36.02
87.5	12	7.29	11	110	36.10
87.75	12	7.31	11	110	36.18
88	12	7.33	11	110	36.25
88.25	12	7.35	11	110	36.33
88.5	12	7.38	11	110	36.41
88.75	12	7.40	11	110	36.49
89	12	7.42	11	110	36.56
89.25	12	7.44	11	110	36.64
89.5	12	7.46	11	110	36.72
89.75	12	7.48	11	110	36.79
90	12	7.50	11	110	36.87
90.25	12	7.52	11	110	36.95
90.5	12	7.54	11	110	37.02
90.75	12	7.56	11	110	37.10
91	12	7.58	11	110	37.17
91.25	13	7.02	12	120	35.07
91.5	13	7.04	12	120	35.14
91.75	13	7.06	12	120	35.21
92	13	7.08	12	120	35.29
92.25	13	7.10	12	120	35.36
92.5	13	7.12	12	120	35.43
92.75	13	7.13	12	120	35.51
93	13	7.15	12	120	35.58
93.25	13	7.17	12	120	35.65
93.5	13	7.19	12	120	35.72
93.75	13	7.21	12	120	35.80
94	13	7.23	12	120	35.87
94.25	13	7.25	12	120	35.94
94.5	13	7.27	12	120	36.01
94.75	13	7.29	12	120	36.09
95	13	7.31	12	120	36.16
95.25	13	7.33	12	120	36.23
95.5	13	7.35	12	120	36.30
95.75	13	7.37	12	120	36.37
96	13	7.38	12	120	36.44
96.25	13	7.40	12	120	36.52
96.5	13	7.42	12	120	36.59
96.75	13	7.44	12	120	36.66

Total Rise	Risers	Riser Height	Steps	Total Run	Angle
97	13	7.46	12	120	36.73
97.25	13	7.48	12	120	36.80
97.5	13	7.50	12	120	36.87
97.75	13	7.52	12	120	36.94
98	14	7.00	13	130	34.99
98.25	14	7.02	13	130	35.06
98.5	14	7.04	13	130	35.13
98.75	14	7.05	13	130	35.20
99	14	7.07	13	130	35.27
99.25	14	7.09	13	130	35.33
99.5	14	7.11	13	130	35.40
99.75	14	7.13	13	130	35.47
100	14	7.14	13	130	35.54
100.25	14	7.16	13	130	35.61
100.5	14	7.18	13	130	35.67
100.75	14	7.20	13	130	35.74
101	14	7.21	13	130	35.81
101.25	14	7.23	13	130	35.87
101.5	14	7.25	13	130	35.94
101.75	14	7.27	13	130	36.01
102	14	7.29	13	130	36.08
102.25	14	7.30	13	130	36.14
102.5	14	7.32	13	130	36.21
102.75	14	7.34	13	130	36.28
103	14	7.36	13	130	36.34
103.25	14	7.38	13	130	36.41
103.5	14	7.39	13	130	36.47
103.75	14	7.41	13	130	36.54
104	14	7.43	13	130	36.61
104.25	14	7.45	13	130	36.67
104.5	14	7.46	13	130	36.74
104.75	14	7.48	13	130	36.80
105	14	7.50	13	130	36.87
105.25	15	7.02	14	140	35.06
105.5	15	7.03	14	140	35.12
105.75	15	7.05	14	140	35.18
106	15	7.07	14	140	35.25
106.25	15	7.08	14	140	35.31
106.5	15	7.10	14	140	35.37
106.75	15	7.12	14	140	35.44
107	15	7.13	14	140	35.50

Total Rise	Risers	Riser Height	Steps	Total Run	Angle
107.25	15	7.15	14	140	35.56
107.5	15	7.17	14	140	35.63
107.75	15	7.18	14	140	35.69
108	15	7.20	14	140	35.75
108.25	15	7.22	14	140	35.82
108.5	15	7.23	14	140	35.88
108.75	15	7.25	14	140	35.94
109	15	7.27	14	140	36.00
109.25	15	7.28	14	140	36.07
109.5	15	7.30	14	140	36.13
109.75	15	7.32	14	140	36.19
110	15	7.33	14	140	36.25
110.25	15	7.35	14	140	36.32
110.5	15	7.37	14	140	36.38
110.75	15	7.38	14	140	36.44
111	15	7.40	14	140	36.50
111.25	15	7.42	14	140	36.56
111.5	15	7.43	14	140	36.62
111.75	15	7.45	14	140	36.69
112	15	7.47	14	140	36.75
112.25	15	7.48	14	140	36.81
112.5	15	7.50	14	140	36.87
112.75	16	7.05	15	150	35.17
113	16	7.06	15	150	35.23
113.25	16	7.08	15	150	35.29
113.5	16	7.09	15	150	35.35
113.75	16	7.11	15	150	35.41
114	16	7.13	15	150	35.47
114.25	16	7.14	15	150	35.53
114.5	16	7.16	15	150	35.59
114.75	16	7.17	15	150	35.65
115	16	7.19	15	150	35.71
115.25	16	7.20	15	150	35.77
115.5	16	7.22	15	150	35.82
115.75	16	7.23	15	150	35.88
116	16	7.25	15	150	35.94
116.25	16	7.27	15	150	36.00
116.5	16	7.28	15	150	36.06
116.75	16	7.30	15	150	36.12
117	16	7.31	15	150	36.18
117.25	16	7.33	15	150	36.23

Total Rise	Risers	Riser Height	Steps	Total Run	Angle
117.5	16	7.34	15	150	36.29
117.75	16	7.36	15	150	36.35
118	16	7.38	15	150	36.41
118.25	16	7.39	15	150	36.47
118.5	16	7.41	15	150	36.52
118.75	16	7.42	15	150	36.58
119	16	7.44	15	150	36.64
119.25	16	7.45	15	150	36.70
119.5	16	7.47	15	150	36.76
119.75	16	7.48	15	150	36.81
120	16	7.50	15	150	36.87
120.25	17	7.07	16	160	35.27
120.5	17	7.09	16	160	35.33
120.75	17	7.10	16	160	35.39
121	17	7.12	16	160	35.44
121.25	17	7.13	16	160	35.50
121.5	17	7.15	16	160	35.55
121.75	17	7.16	16	160	35.61
122	17	7.18	16	160	35.67
122.25	17	7.19	16	160	35.72
122.5	17	7.21	16	160	35.78
122.75	17	7.22	16	160	35.83
123	17	7.24	16	160	35.89
123.25	17	7.25	16	160	35.94
123.5	17	7.26	16	160	36.00
123.75	17	7.28	16	160	36.05
124	17	7.29	16	160	36.11
124.25	17	7.31	16	160	36.16
124.5	17	7.32	16	160	36.22
124.75	17	7.34	16	160	36.27
125	17	7.35	16	160	36.33
125.25	17	7.37	16	160	36.38
125.5	17	7.38	16	160	36.44
125.75	17	7.40	16	160	36.49
126	17	7.41	16	160	36.54
126.25	17	7.43	16	160	36.60
126.5	17	7.44	16	160	36.65
126.75	17	7.46	16	160	36.71
127	17	7.47	16	160	36.76
127.25	17	7.49	16	160	36.82
127.5	17	7.50	16	160	36.87

Total Rise	Risers	Riser Height	Steps	Total Run	Angle
127.75	18	7.10	17	170	35.36
128	18	7.11	17	170	35.42
128.25	18	7.13	17	170	35.47
128.5	18	7.14	17	170	35.52
128.75	18	7.15	17	170	35.58
129	18	7.17	17	170	35.63
129.25	18	7.18	17	170	35.68
129.5	18	7.19	17	170	35.73
129.75	18	7.21	17	170	35.79
130	18	7.22	17	170	35.84
130.25	18	7.24	17	170	35.89
130.5	18	7.25	17	170	35.94
130.75	18	7.26	17	170	35.99
131	18	7.28	17	170	36.05
131.25	18	7.29	17	170	36.10
131.5	18	7.31	17	170	36.15
131.75	18	7.32	17	170	36.20
132	18	7.33	17	170	36.25
132.25	18	7.35	17	170	36.31
132.5	18	7.36	17	170	36.36
132.75	18	7.38	17	170	36.41
133	18	7.39	17	170	36.46
133.25	18	7.40	17	170	36.51
133.5	18	7.42	17	170	36.56
133.75	18	7.43	17	170	36.61
134	18	7.44	17	170	36.67
134.25	18	7.46	17	170	36.72
134.5	18	7.47	17	170	36.77
134.75	18	7.49	17	170	36.82
135	18	7.50	17	170	36.87
135.25	19	7.12	18	180	35.44
135.5	19	7.13	18	180	35.49
135.75	19	7.14	18	180	35.54
135	19	7.11	18	180	35.39
135.25	19	7.12	18	180	35.44
135.5	19	7.13	18	180	35.49
135.75	19	7.14	18	180	35.54
136	19	7.16	18	180	35.59
136.25	19	7.17	18	180	35.64
136.5	19	7.18	18	180	35.69
136.75	19	7.20	18	180	35.74

Total Rise	Risers	Riser Height	Steps	Total Run	Angle
137	19	7.21	18	180	35.79
137.25	19	7.22	18	180	35.84
137.5	19	7.24	18	180	35.89
137.75	19	7.25	18	180	35.94
138	19	7.26	18	180	35.99
138.25	19	7.28	18	180	36.04
138.5	19	7.29	18	180	36.09
138.75	19	7.30	18	180	36.14
139	19	7.32	18	180	36.19
139.25	19	7.33	18	180	36.24
139.5	19	7.34	18	180	36.29
139.75	19	7.36	18	180	36.34
140	19	7.37	18	180	36.38

Rise And Run Chart For 17 – ½ Inch Rule

Total Rise = Total height of stairway, from the top of the bottom floor to the top of the top floor.

Risers = Amount of risers in stairway.

Riser Height = Individual riser height or total height between steps.

Steps = Amount of steps in stairway.

Total Run = Length of stairway in inches, while subtracting height of riser from 17.5 inches.

Angle = This is the angle of the stairway or incline.

Total Rise	Risers	Riser Height	Steps	Tread	Total Run	Angle
10	2	5.00	1	12.50	12.50	21.80
10.25	2	5.13	1	12.38	12.38	22.50
10.5	2	5.25	1	12.25	12.25	23.20
10.75	2	5.38	1	12.13	12.13	23.91
10.75	2	5.38	1	12.13	12.13	23.91
11.25	2	5.63	1	11.88	11.88	25.35
11.5	2	5.75	1	11.75	11.75	26.08
11.75	2	5.88	1	11.63	11.63	26.81
12	2	6.00	1	11.50	11.50	27.55
12.25	2	6.13	1	11.38	11.38	28.30
12.5	2	6.25	1	11.25	11.25	29.05
12.75	2	6.38	1	11.13	11.13	29.81
13	2	6.50	1	11.00	11.00	30.58
13.25	2	6.63	1	10.88	10.88	31.35
13.5	2	6.75	1	10.75	10.75	32.12
13.75	2	6.88	1	10.63	10.63	32.91
14	2	7.00	1	10.50	10.50	33.69
14.25	2	7.13	1	10.38	10.38	34.48

Total Rise	Risers	Riser Height	Steps	Tread	Total Run	Angle
14.5	2	7.25	1	10.25	10.25	35.27
14.75	2	7.38	1	10.13	10.13	36.07
15	2	7.50	1	10.00	10.00	36.87
15.25	2	7.63	1	9.88	9.88	37.67
15.5	2	7.75	1	9.75	9.75	38.48
15.75	3	5.25	2	12.25	24.50	23.20
16	3	5.33	2	12.17	24.33	23.67
16.25	3	5.42	2	12.08	24.17	24.15
16.5	3	5.50	2	12.00	24.00	24.62
16.75	3	5.58	2	11.92	23.83	25.10
17	3	5.67	2	11.83	23.67	25.59
17.25	3	5.75	2	11.75	23.50	26.08
17.5	3	5.83	2	11.67	23.33	26.57
17.75	3	5.92	2	11.58	23.17	27.06
18	3	6.00	2	11.50	23.00	27.55
18.25	3	6.08	2	11.42	22.83	28.05
18.5	3	6.17	2	11.33	22.67	28.55
18.75	3	6.25	2	11.25	22.50	29.05
19	3	6.33	2	11.17	22.33	29.56
19.25	3	6.42	2	11.08	22.17	30.07
19.5	3	6.50	2	11.00	22.00	30.58
19.75	3	6.58	2	10.92	21.83	31.09
20	3	6.67	2	10.83	21.67	31.61
20.25	3	6.75	2	10.75	21.50	32.12
20.5	3	6.83	2	10.67	21.33	32.64
20.75	3	6.92	2	10.58	21.17	33.17
21	3	7.00	2	10.50	21.00	33.69
21.25	3	7.08	2	10.42	20.83	34.22
21.5	3	7.17	2	10.33	20.67	34.74
21.75	3	7.25	2	10.25	20.50	35.27
22	3	7.33	2	10.17	20.33	35.80
22.25	3	7.42	2	10.08	20.17	36.34
22.5	3	7.50	2	10.00	20.00	36.87
22.75	3	7.58	2	9.92	19.83	37.41
23	3	7.67	2	9.83	19.67	37.94
23.25	3	7.75	2	9.75	19.50	38.48
23.5	4	5.88	3	11.63	34.88	26.81
23.75	4	5.94	3	11.56	34.69	27.18
24	4	6.00	3	11.50	34.50	27.55

Total Rise	Risers	Riser Height	Steps	Tread	Total Run	Angle
24.25	4	6.06	3	11.44	34.31	27.93
24.5	4	6.13	3	11.38	34.13	28.30
24.75	4	6.19	3	11.31	33.94	28.68
25	4	6.25	3	11.25	33.75	29.05
25.25	4	6.31	3	11.19	33.56	29.43
25.5	4	6.38	3	11.13	33.38	29.81
25.75	4	6.44	3	11.06	33.19	30.20
26	4	6.50	3	11.00	33.00	30.58
26.25	4	6.56	3	10.94	32.81	30.96
26.5	4	6.63	3	10.88	32.63	31.35
26.75	4	6.69	3	10.81	32.44	31.74
27	4	6.75	3	10.75	32.25	32.12
27.25	4	6.81	3	10.69	32.06	32.51
27.5	4	6.88	3	10.63	31.88	32.91
27.75	4	6.94	3	10.56	31.69	33.30
28	4	7.00	3	10.50	31.50	33.69
28.25	4	7.06	3	10.44	31.31	34.08
28.5	4	7.13	3	10.38	31.13	34.48
28.75	4	7.19	3	10.31	30.94	34.88
29	4	7.25	3	10.25	30.75	35.27
29.25	4	7.31	3	10.19	30.56	35.67
29.5	4	7.38	3	10.13	30.38	36.07
29.75	4	7.44	3	10.06	30.19	36.47
30	4	7.50	3	10.00	30.00	36.87
30.25	4	7.56	3	9.94	29.81	37.27
30.5	4	7.63	3	9.88	29.63	37.67
30.75	4	7.69	3	9.81	29.44	38.08
31	4	7.75	3	9.75	29.25	38.48
31.25	5	6.25	4	11.25	45.00	29.05
31.5	5	6.30	4	11.20	44.80	29.36
31.75	5	6.35	4	11.15	44.60	29.66
32	5	6.40	4	11.10	44.40	29.97
32.25	5	6.45	4	11.05	44.20	30.27
32.5	5	6.50	4	11.00	44.00	30.58
32.75	5	6.55	4	10.95	43.80	30.89
33	5	6.60	4	10.90	43.60	31.20
33.25	5	6.65	4	10.85	43.40	31.50
33.5	5	6.70	4	10.80	43.20	31.81
33.75	5	6.75	4	10.75	43.00	32.12

Total Rise	Risers	Riser Height	Steps	Tread	Total Run	Angle
34	5	6.80	4	10.70	42.80	32.44
34.25	5	6.85	4	10.65	42.60	32.75
34.5	5	6.90	4	10.60	42.40	33.06
34.75	5	6.95	4	10.55	42.20	33.38
35	5	7.00	4	10.50	42.00	33.69
35.25	5	7.05	4	10.45	41.80	34.01
35.5	5	7.10	4	10.40	41.60	34.32
35.75	5	7.15	4	10.35	41.40	34.64
36	5	7.20	4	10.30	41.20	34.95
36.25	5	7.25	4	10.25	41.00	35.27
36.5	5	7.30	4	10.20	40.80	35.59
36.75	5	7.35	4	10.15	40.60	35.91
37	5	7.40	4	10.10	40.40	36.23
37.25	5	7.45	4	10.05	40.20	36.55
37.5	5	7.50	4	10.00	40.00	36.87
37.75	5	7.55	4	9.95	39.80	37.19
38	5	7.60	4	9.90	39.60	37.51
38.25	5	7.65	4	9.85	39.40	37.83
38.5	5	7.70	4	9.80	39.20	38.16
38.75	5	7.75	4	9.75	39.00	38.48
39	6	6.50	5	11.00	55.00	30.58
39.25	6	6.54	5	10.96	54.79	30.84
39.5	6	6.58	5	10.92	54.58	31.09
39.75	6	6.63	5	10.88	54.38	31.35
40	6	6.67	5	10.83	54.17	31.61
40.25	6	6.71	5	10.79	53.96	31.87
40.5	6	6.75	5	10.75	53.75	32.12
40.75	6	6.79	5	10.71	53.54	32.38
41	6	6.83	5	10.67	53.33	32.64
41.25	6	6.88	5	10.63	53.13	32.91
41.5	6	6.92	5	10.58	52.92	33.17
41.75	6	6.96	5	10.54	52.71	33.43
42	6	7.00	5	10.50	52.50	33.69
42.25	6	7.04	5	10.46	52.29	33.95
42.5	6	7.08	5	10.42	52.08	34.22
42.75	6	7.13	5	10.38	51.88	34.48
43	6	7.17	5	10.33	51.67	34.74
43.25	6	7.21	5	10.29	51.46	35.01
43.5	6	7.25	5	10.25	51.25	35.27

Total Rise	Risers	Riser Height	Steps	Tread	Total Run	Angle
43.75	6	7.29	5	10.21	51.04	35.54
44	6	7.33	5	10.17	50.83	35.80
44.25	6	7.38	5	10.13	50.63	36.07
44.5	6	7.42	5	10.08	50.42	36.34
44.75	6	7.46	5	10.04	50.21	36.60
45	6	7.50	5	10.00	50.00	36.87
45.25	6	7.54	5	9.96	49.79	37.14
45.5	6	7.58	5	9.92	49.58	37.41
45.75	6	7.63	5	9.88	49.38	37.67
46	6	7.67	5	9.83	49.17	37.94
46.25	6	7.71	5	9.79	48.96	38.21
46.5	6	7.75	5	9.75	48.75	38.48
46.75	7	6.68	6	10.82	64.93	31.68
47	7	6.71	6	10.79	64.71	31.90
47.25	7	6.75	6	10.75	64.50	32.12
47.5	7	6.79	6	10.71	64.29	32.35
47.75	7	6.82	6	10.68	64.07	32.57
48	7	6.86	6	10.64	63.86	32.79
48.25	7	6.89	6	10.61	63.64	33.02
48.5	7	6.93	6	10.57	63.43	33.24
48.75	7	6.96	6	10.54	63.21	33.47
49	7	7.00	6	10.50	63.00	33.69
49.25	7	7.04	6	10.46	62.79	33.92
49.5	7	7.07	6	10.43	62.57	34.14
49.75	7	7.11	6	10.39	62.36	34.37
50	7	7.14	6	10.36	62.14	34.59
5.25	7	0.75	6	16.75	100.50	2.56
50.5	7	7.21	6	10.29	61.71	35.05
50.75	7	7.25	6	10.25	61.50	35.27
51	7	7.29	6	10.21	61.29	35.50
51.25	7	7.32	6	10.18	61.07	35.73
51.5	7	7.36	6	10.14	60.86	35.96
51.75	7	7.39	6	10.11	60.64	36.18
52	7	7.43	6	10.07	60.43	36.41
52.25	7	7.46	6	10.04	60.21	36.64
52.5	7	7.50	6	10.00	60.00	36.87
52.75	7	7.54	6	9.96	59.79	37.10
53	7	7.57	6	9.93	59.57	37.33
53.25	7	7.61	6	9.89	59.36	37.56

Total Rise	Risers	Riser Height	Steps	Tread	Total Run	Angle
53.5	7	7.64	6	9.86	59.14	37.79
53.75	7	7.68	6	9.82	58.93	38.02
54	7	7.71	6	9.79	58.71	38.25
54.25	7	7.75	6	9.75	58.50	38.48
54.5	8	6.81	7	10.69	74.81	32.51
54.75	8	6.84	7	10.66	74.59	32.71
55	8	6.88	7	10.63	74.38	32.91
55.25	8	6.91	7	10.59	74.16	33.10
55.5	8	6.94	7	10.56	73.94	33.30
55.75	8	6.97	7	10.53	73.72	33.49
56	8	7.00	7	10.50	73.50	33.69
56.25	8	7.03	7	10.47	73.28	33.89
56.5	8	7.06	7	10.44	73.06	34.08
56.75	8	7.09	7	10.41	72.84	34.28
57	8	7.13	7	10.38	72.63	34.48
57.25	8	7.16	7	10.34	72.41	34.68
57.5	8	7.19	7	10.31	72.19	34.88
57.75	8	7.22	7	10.28	71.97	35.07
58	8	7.25	7	10.25	71.75	35.27
58.25	8	7.28	7	10.22	71.53	35.47
58.5	8	7.31	7	10.19	71.31	35.67
58.75	8	7.34	7	10.16	71.09	35.87
59	8	7.38	7	10.13	70.88	36.07
59.25	8	7.41	7	10.09	70.66	36.27
59.5	8	7.44	7	10.06	70.44	36.47
59.75	8	7.47	7	10.03	70.22	36.67
60	8	7.50	7	10.00	70.00	36.87
60.25	8	7.53	7	9.97	69.78	37.07
60.5	8	7.56	7	9.94	69.56	37.27
60.75	8	7.59	7	9.91	69.34	37.47
61	8	7.63	7	9.88	69.13	37.67
61.25	8	7.66	7	9.84	68.91	37.87
61.5	8	7.69	7	9.81	68.69	38.08
61.75	8	7.72	7	9.78	68.47	38.28
62	8	7.75	7	9.75	68.25	38.48
62.25	9	6.92	8	10.58	84.67	33.17
62.5	9	6.94	8	10.56	84.44	33.34
62.75	9	6.97	8	10.53	84.22	33.52
63	9	7.00	8	10.50	84.00	33.69

Total Rise	Risers	Riser Height	Steps	Tread	Total Run	Angle
63.25	9	7.03	8	10.47	83.78	33.87
63.5	9	7.06	8	10.44	83.56	34.04
63.75	9	7.08	8	10.42	83.33	34.22
64	9	7.11	8	10.39	83.11	34.39
64.25	9	7.14	8	10.36	82.89	34.57
64.5	9	7.17	8	10.33	82.67	34.74
64.75	9	7.19	8	10.31	82.44	34.92
65	9	7.22	8	10.28	82.22	35.10
65.25	9	7.25	8	10.25	82.00	35.27
65.5	9	7.28	8	10.22	81.78	35.45
65.75	9	7.31	8	10.19	81.56	35.63
66	9	7.33	8	10.17	81.33	35.80
66.25	9	7.36	8	10.14	81.11	35.98
66.5	9	7.39	8	10.11	80.89	36.16
66.75	9	7.42	8	10.08	80.67	36.34
67	9	7.44	8	10.06	80.44	36.51
67.25	9	7.47	8	10.03	80.22	36.69
67.5	9	7.50	8	10.00	80.00	36.87
67.75	9	7.53	8	9.97	79.78	37.05
68	9	7.56	8	9.94	79.56	37.23
68.25	9	7.58	8	9.92	79.33	37.41
68.5	9	7.61	8	9.89	79.11	37.58
68.75	9	7.64	8	9.86	78.89	37.76
69	10	6.90	9	10.60	95.40	33.06
69.25	10	6.93	9	10.58	95.18	33.22
69.5	10	6.95	9	10.55	94.95	33.38
69.75	10	6.98	9	10.53	94.73	33.53
70	10	7.00	9	10.50	94.50	33.69
70.25	10	7.03	9	10.48	94.28	33.85
70.5	10	7.05	9	10.45	94.05	34.01
70.75	10	7.08	9	10.43	93.83	34.16
71	10	7.10	9	10.40	93.60	34.32
71.25	10	7.13	9	10.38	93.38	34.48
71.5	10	7.15	9	10.35	93.15	34.64
71.75	10	7.18	9	10.33	92.93	34.80
72	10	7.20	9	10.30	92.70	34.95
72.25	10	7.23	9	10.28	92.48	35.11
72.5	10	7.25	9	10.25	92.25	35.27
72.75	10	7.28	9	10.23	92.03	35.43

Total Rise	Risers	Riser Height	Steps	Tread	Total Run	Angle
73	10	7.30	9	10.20	91.80	35.59
73.25	10	7.33	9	10.18	91.58	35.75
73.5	10	7.35	9	10.15	91.35	35.91
73.75	10	7.38	9	10.13	91.13	36.07
74	10	7.40	9	10.10	90.90	36.23
74.25	10	7.43	9	10.08	90.68	36.39
74.5	10	7.45	9	10.05	90.45	36.55
74.75	10	7.48	9	10.03	90.23	36.71
75	10	7.50	9	10.00	90.00	36.87
75.25	10	7.53	9	9.98	89.78	37.03
75.5	10	7.55	9	9.95	89.55	37.19
75.75	10	7.58	9	9.93	89.33	37.35
76	10	7.60	9	9.90	89.10	37.51
76.25	10	7.63	9	9.88	88.88	37.67
76.5	11	6.95	10	10.55	105.45	33.40
76.75	11	6.98	10	10.52	105.23	33.55
77	11	7.00	10	10.50	105.00	33.69
77.25	11	7.02	10	10.48	104.77	33.83
77.5	11	7.05	10	10.45	104.55	33.98
77.75	11	7.07	10	10.43	104.32	34.12
78	11	7.09	10	10.41	104.09	34.26
78.25	11	7.11	10	10.39	103.86	34.41
78.5	11	7.14	10	10.36	103.64	34.55
78.75	11	7.16	10	10.34	103.41	34.70
79	11	7.18	10	10.32	103.18	34.84
79.25	11	7.20	10	10.30	102.95	34.98
79.5	11	7.23	10	10.27	102.73	35.13
79.75	11	7.25	10	10.25	102.50	35.27
80	11	7.27	10	10.23	102.27	35.42
80.25	11	7.30	10	10.20	102.05	35.56
80.5	11	7.32	10	10.18	101.82	35.71
80.75	11	7.34	10	10.16	101.59	35.85
81	11	7.36	10	10.14	101.36	36.00
81.25	11	7.39	10	10.11	101.14	36.14
81.5	11	7.41	10	10.09	100.91	36.29
81.75	11	7.43	10	10.07	100.68	36.43
82	11	7.45	10	10.05	100.45	36.58
82.25	11	7.48	10	10.02	100.23	36.72
82.5	11	7.50	10	10.00	100.00	36.87

Total Rise	Risers	Riser Height	Steps	Tread	Total Run	Angle
82.75	11	7.52	10	9.98	99.77	37.02
83	11	7.55	10	9.95	99.55	37.16
83.25	11	7.57	10	9.93	99.32	37.31
83.5	11	7.59	10	9.91	99.09	37.45
83.75	11	7.61	10	9.89	98.86	37.60
84	11	7.64	10	9.86	98.64	37.75
84.25	12	7.02	11	10.48	115.27	33.82
84.5	12	7.04	11	10.46	115.04	33.95
84.75	12	7.06	11	10.44	114.81	34.08
85	12	7.08	11	10.42	114.58	34.22
85.25	12	7.10	11	10.40	114.35	34.35
85.5	12	7.13	11	10.38	114.13	34.48
85.75	12	7.15	11	10.35	113.90	34.61
86	12	7.17	11	10.33	113.67	34.74
86.25	12	7.19	11	10.31	113.44	34.88
86.5	12	7.21	11	10.29	113.21	35.01
86.75	12	7.23	11	10.27	112.98	35.14
87	12	7.25	11	10.25	112.75	35.27
87.25	12	7.27	11	10.23	112.52	35.40
87.5	12	7.29	11	10.21	112.29	35.54
87.75	12	7.31	11	10.19	112.06	35.67
88	12	7.33	11	10.17	111.83	35.80
88.25	12	7.35	11	10.15	111.60	35.94
88.5	12	7.38	11	10.13	111.38	36.07
88.75	12	7.40	11	10.10	111.15	36.20
89	12	7.42	11	10.08	110.92	36.34
89.25	12	7.44	11	10.06	110.69	36.47
89.5	12	7.46	11	10.04	110.46	36.60
89.75	12	7.48	11	10.02	110.23	36.74
90	12	7.50	11	10.00	110.00	36.87
90.25	12	7.52	11	9.98	109.77	37.00
90.5	12	7.54	11	9.96	109.54	37.14
90.75	12	7.56	11	9.94	109.31	37.27
91	12	7.58	11	9.92	109.08	37.41
91.25	13	7.02	12	10.48	125.77	33.81
91.5	13	7.04	12	10.46	125.54	33.93
91.75	13	7.06	12	10.44	125.31	34.05
92	13	7.08	12	10.42	125.08	34.18
92.25	13	7.10	12	10.40	124.85	34.30

Total Rise	Risers	Riser Height	Steps	Tread	Total Run	Angle
92.5	13	7.12	12	10.38	124.62	34.42
92.75	13	7.13	12	10.37	124.38	34.54
93	13	7.15	12	10.35	124.15	34.66
93.25	13	7.17	12	10.33	123.92	34.78
93.5	13	7.19	12	10.31	123.69	34.91
93.75	13	7.21	12	10.29	123.46	35.03
94	13	7.23	12	10.27	123.23	35.15
94.25	13	7.25	12	10.25	123.00	35.27
94.5	13	7.27	12	10.23	122.77	35.39
94.75	13	7.29	12	10.21	122.54	35.52
95	13	7.31	12	10.19	122.31	35.64
95.25	13	7.33	12	10.17	122.08	35.76
95.5	13	7.35	12	10.15	121.85	35.89
95.75	13	7.37	12	10.13	121.62	36.01
96	13	7.38	12	10.12	121.38	36.13
96.25	13	7.40	12	10.10	121.15	36.25
96.5	13	7.42	12	10.08	120.92	36.38
96.75	13	7.44	12	10.06	120.69	36.50
97	13	7.46	12	10.04	120.46	36.62
97.25	13	7.48	12	10.02	120.23	36.75
97.5	13	7.50	12	10.00	120.00	36.87
97.75	13	7.52	12	9.98	119.77	36.99
98	14	7.00	13	10.50	136.50	33.69
98.25	14	7.02	13	10.48	136.27	33.80
98.5	14	7.04	13	10.46	136.04	33.92
98.75	14	7.05	13	10.45	135.80	34.03
99	14	7.07	13	10.43	135.57	34.14
99.25	14	7.09	13	10.41	135.34	34.25
99.5	14	7.11	13	10.39	135.11	34.37
99.75	14	7.13	13	10.38	134.88	34.48
100	14	7.14	13	10.36	134.64	34.59
100.25	14	7.16	13	10.34	134.41	34.71
100.5	14	7.18	13	10.32	134.18	34.82
100.75	14	7.20	13	10.30	133.95	34.93
101	14	7.21	13	10.29	133.71	35.05
101.25	14	7.23	13	10.27	133.48	35.16
101.5	14	7.25	13	10.25	133.25	35.27
101.75	14	7.27	13	10.23	133.02	35.39
102	14	7.29	13	10.21	132.79	35.50

Total Rise	Risers	Riser Height	Steps	Tread	Total Run	Angle
102.25	14	7.30	13	10.20	132.55	35.61
102.5	14	7.32	13	10.18	132.32	35.73
102.75	14	7.34	13	10.16	132.09	35.84
103	14	7.36	13	10.14	131.86	35.96
103.25	14	7.38	13	10.13	131.63	36.07
103.5	14	7.39	13	10.11	131.39	36.18
103.75	14	7.41	13	10.09	131.16	36.30
104	14	7.43	13	10.07	130.93	36.41
104.25	14	7.45	13	10.05	130.70	36.53
104.5	14	7.46	13	10.04	130.46	36.64
104.75	14	7.48	13	10.02	130.23	36.76
105	14	7.50	13	10.00	130.00	36.87
105.25	15	7.02	14	10.48	146.77	33.80
105.5	15	7.03	14	10.47	146.53	33.90
105.75	15	7.05	14	10.45	146.30	34.01
106	15	7.07	14	10.43	146.07	34.11
106.25	15	7.08	14	10.42	145.83	34.22
106.5	15	7.10	14	10.40	145.60	34.32
106.75	15	7.12	14	10.38	145.37	34.43
107	15	7.13	14	10.37	145.13	34.53
107.25	15	7.15	14	10.35	144.90	34.64
107.5	15	7.17	14	10.33	144.67	34.74
107.75	15	7.18	14	10.32	144.43	34.85
108	15	7.20	14	10.30	144.20	34.95
108.25	15	7.22	14	10.28	143.97	35.06
108.5	15	7.23	14	10.27	143.73	35.17
108.75	15	7.25	14	10.25	143.50	35.27
109	15	7.27	14	10.23	143.27	35.38
109.25	15	7.28	14	10.22	143.03	35.48
109.5	15	7.30	14	10.20	142.80	35.59
109.75	15	7.32	14	10.18	142.57	35.70
110	15	7.33	14	10.17	142.33	35.80
110.25	15	7.35	14	10.15	142.10	35.91
110.5	15	7.37	14	10.13	141.87	36.02
110.75	15	7.38	14	10.12	141.63	36.12
111	15	7.40	14	10.10	141.40	36.23
111.25	15	7.42	14	10.08	141.17	36.34
111.5	15	7.43	14	10.07	140.93	36.44
111.75	15	7.45	14	10.05	140.70	36.55

Total Rise	Risers	Riser Height	Steps	Tread	Total Run	Angle
112	15	7.47	14	10.03	140.47	36.66
112.25	15	7.48	14	10.02	140.23	36.76
112.5	15	7.50	14	10.00	140.00	36.87
112.75	16	7.05	15	10.45	156.80	33.99
113	16	7.06	15	10.44	156.56	34.08
113.25	16	7.08	15	10.42	156.33	34.18
113.5	16	7.09	15	10.41	156.09	34.28
113.75	16	7.11	15	10.39	155.86	34.38
114	16	7.13	15	10.38	155.63	34.48
114.25	16	7.14	15	10.36	155.39	34.58
114.5	16	7.16	15	10.34	155.16	34.68
114.75	16	7.17	15	10.33	154.92	34.78
115	16	7.19	15	10.31	154.69	34.88
115.25	16	7.20	15	10.30	154.45	34.97
115.5	16	7.22	15	10.28	154.22	35.07
115.75	16	7.23	15	10.27	153.98	35.17
116	16	7.25	15	10.25	153.75	35.27
116.25	16	7.27	15	10.23	153.52	35.37
116.5	16	7.28	15	10.22	153.28	35.47
116.75	16	7.30	15	10.20	153.05	35.57
117	16	7.31	15	10.19	152.81	35.67
117.25	16	7.33	15	10.17	152.58	35.77
117.5	16	7.34	15	10.16	152.34	35.87
117.75	16	7.36	15	10.14	152.11	35.97
118	16	7.38	15	10.13	151.88	36.07
118.25	16	7.39	15	10.11	151.64	36.17
118.5	16	7.41	15	10.09	151.41	36.27
118.75	16	7.42	15	10.08	151.17	36.37
119	16	7.44	15	10.06	150.94	36.47
119.25	16	7.45	15	10.05	150.70	36.57
119.5	16	7.47	15	10.03	150.47	36.67
119.75	16	7.48	15	10.02	150.23	36.77
120	16	7.50	15	10.00	150.00	36.87
120.25	17	7.07	16	10.43	166.82	34.15
120.5	17	7.09	16	10.41	166.59	34.25
120.75	17	7.10	16	10.40	166.35	34.34
121	17	7.12	16	10.38	166.12	34.43
121.25	17	7.13	16	10.37	165.88	34.53
121.5	17	7.15	16	10.35	165.65	34.62

Total Rise	Risers	Riser Height	Steps	Tread	Total Run	Angle
121.75	17	7.16	16	10.34	165.41	34.71
122	17	7.18	16	10.32	165.18	34.81
122.25	17	7.19	16	10.31	164.94	34.90
122.5	17	7.21	16	10.29	164.71	34.99
122.75	17	7.22	16	10.28	164.47	35.09
123	17	7.24	16	10.26	164.24	35.18
123.25	17	7.25	16	10.25	164.00	35.27
123.5	17	7.26	16	10.24	163.76	35.37
123.75	17	7.28	16	10.22	163.53	35.46
124	17	7.29	16	10.21	163.29	35.55
124.25	17	7.31	16	10.19	163.06	35.65
124.5	17	7.32	16	10.18	162.82	35.74
124.75	17	7.34	16	10.16	162.59	35.83
125	17	7.35	16	10.15	162.35	35.93
125.25	17	7.37	16	10.13	162.12	36.02
125.5	17	7.38	16	10.12	161.88	36.12
125.75	17	7.40	16	10.10	161.65	36.21
126	17	7.41	16	10.09	161.41	36.30
126.25	17	7.43	16	10.07	161.18	36.40
126.5	17	7.44	16	10.06	160.94	36.49
126.75	17	7.46	16	10.04	160.71	36.59
127	17	7.47	16	10.03	160.47	36.68
127.25	17	7.49	16	10.01	160.24	36.78
127.5	17	7.50	16	10.00	160.00	36.87
127.75	18	7.10	17	10.40	176.85	34.30
128	18	7.11	17	10.39	176.61	34.39
128.25	18	7.13	17	10.38	176.38	34.48
128.5	18	7.14	17	10.36	176.14	34.57
128.75	18	7.15	17	10.35	175.90	34.66
129	18	7.17	17	10.33	175.67	34.74
129.25	18	7.18	17	10.32	175.43	34.83
129.5	18	7.19	17	10.31	175.19	34.92
129.75	18	7.21	17	10.29	174.96	35.01
130	18	7.22	17	10.28	174.72	35.10
130.25	18	7.24	17	10.26	174.49	35.18
130.5	18	7.25	17	10.25	174.25	35.27
130.75	18	7.26	17	10.24	174.01	35.36
131	18	7.28	17	10.22	173.78	35.45
131.25	18	7.29	17	10.21	173.54	35.54

Total Rise	Risers	Riser Height	Steps	Tread	Total Run	Angle
131.5	18	7.31	17	10.19	173.31	35.63
131.75	18	7.32	17	10.18	173.07	35.71
132	18	7.33	17	10.17	172.83	35.80
132.25	18	7.35	17	10.15	172.60	35.89
132.5	18	7.36	17	10.14	172.36	35.98
132.75	18	7.38	17	10.13	172.13	36.07
133	18	7.39	17	10.11	171.89	36.16
133.25	18	7.40	17	10.10	171.65	36.25
133.5	18	7.42	17	10.08	171.42	36.34
133.75	18	7.43	17	10.07	171.18	36.42
134	18	7.44	17	10.06	170.94	36.51
134.25	18	7.46	17	10.04	170.71	36.60
134.5	18	7.47	17	10.03	170.47	36.69
134.75	18	7.49	17	10.01	170.24	36.78
135	18	7.50	17	10.00	170.00	36.87
135.25	19	7.12	18	10.38	186.87	34.44
135.5	19	7.13	18	10.37	186.63	34.52
135.75	19	7.14	18	10.36	186.39	34.60
135	19	7.11	18	10.39	187.11	34.35
135.25	19	7.12	18	10.38	186.87	34.44
135.5	19	7.13	18	10.37	186.63	34.52
135.75	19	7.14	18	10.36	186.39	34.60
136	19	7.16	18	10.34	186.16	34.69
136.25	19	7.17	18	10.33	185.92	34.77
136.5	19	7.18	18	10.32	185.68	34.85
136.75	19	7.20	18	10.30	185.45	34.94
137	19	7.21	18	10.29	185.21	35.02
137.25	19	7.22	18	10.28	184.97	35.11
137.5	19	7.24	18	10.26	184.74	35.19
137.75	19	7.25	18	10.25	184.50	35.27
138	19	7.26	18	10.24	184.26	35.36
138.25	19	7.28	18	10.22	184.03	35.44
138.5	19	7.29	18	10.21	183.79	35.52
138.75	19	7.30	18	10.20	183.55	35.61
139	19	7.32	18	10.18	183.32	35.69
139.25	19	7.33	18	10.17	183.08	35.78
139.5	19	7.34	18	10.16	182.84	35.86
139.75	19	7.36	18	10.14	182.61	35.94
140	19	7.37	18	10.13	182.37	36.03

Decimals To Inches Chart

One slash represents a foot (')or 12' translates into twelve feet.

Two slashes or quotation marks represent inches (") or 14" translates into fourteen inches.

These marks are usually used by architects and designers and can be found on building blueprints.

Decimal	Fraction
.0625	One sixteenth of an inch or 1/16"
.125	One eight of an inch or 1/8"
.1875	Three sixteenths of an inch or 3/16"
.25	One quarter inch or 1/4"
.3125	Five sixteenths of and inch or 5/16"
.375	Three eights of an inch or 3/8"
.4375	Seven sixteenths of an inch or 7/16"
.5	One half inch or 1/2"
.5625	Nine sixteenths of an inch or 9/16"
.625	Five eights of an inch or 5/8"
.6875	Eleven sixteenths of an inch or 11/16"
.75	Three quarters of an inch or 3/4"
.8125	Thirteen sixteenths of an inch or 13/16"
.875	Seven eights of an inch or 7/8"
.9375	Fifteen sixteenths of an inch or 15/16"
1.0	One inch or 1"

Winder Basics

I wasn't going to share this information, but decided to when someone I care about dearly suggested otherwise. I built this type of stairway in one of my homes and either witnessed it myself or heard about it later as everyone in our family stumbled and fell down the stairs while walking down, but never fell while walking up.

No one was seriously injured, but it definitely got me thinking about whether or not this type of stairway is safe.

If I could have done it over again, I would've built the stairs with a flat landing and actually ended up doing so later when I built another room addition on the other side of the home.

However, the codes have changed since then and now require wider steps, making them much safer. One particular building code that might be used by your local building department requires an 11 inch minimum step, 12 inches from the inside of the stairway (illustration on page 95).

This is only a suggestion, but if you can build a stairway without this type of design, it might save you or the property owners some grief in the future.

Check with your local building department for more information.

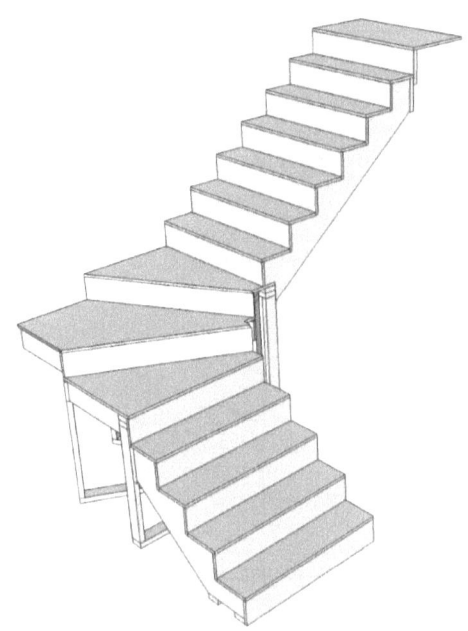

Most winder type stairways have three angled steps like the illustration above and below, but can have more.

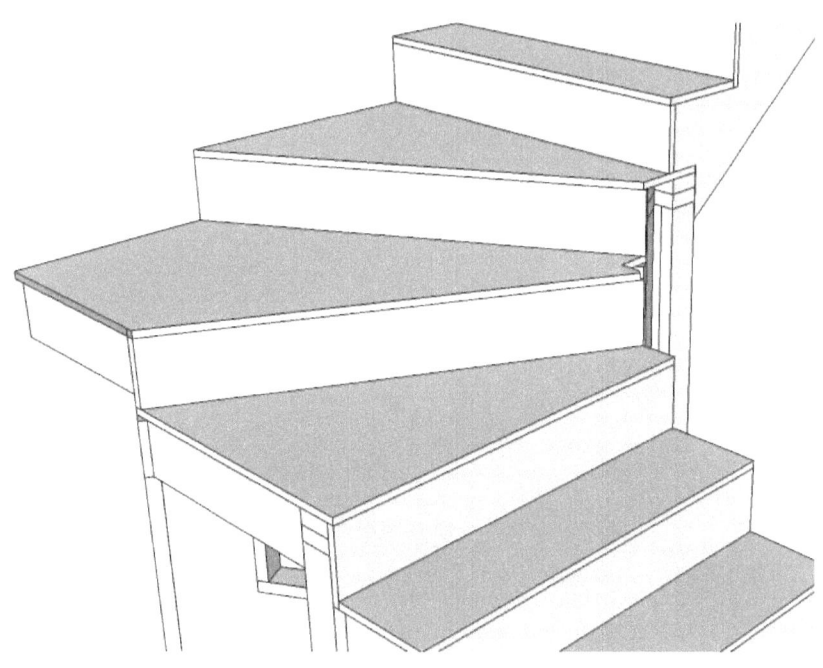

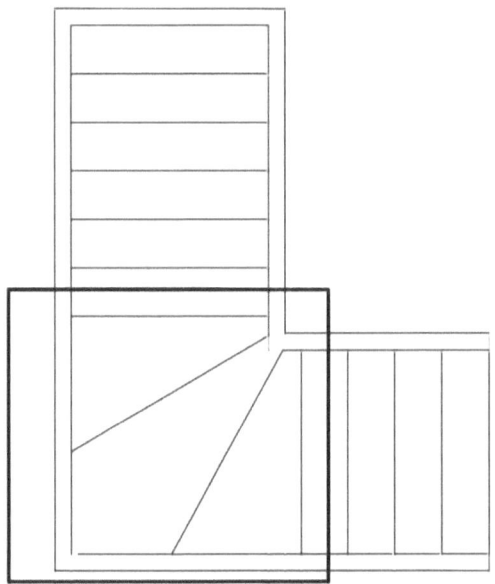

Here are a couple of three step winder floor plans with the illustration below providing you with a building code walk line and an example of what some building codes require.

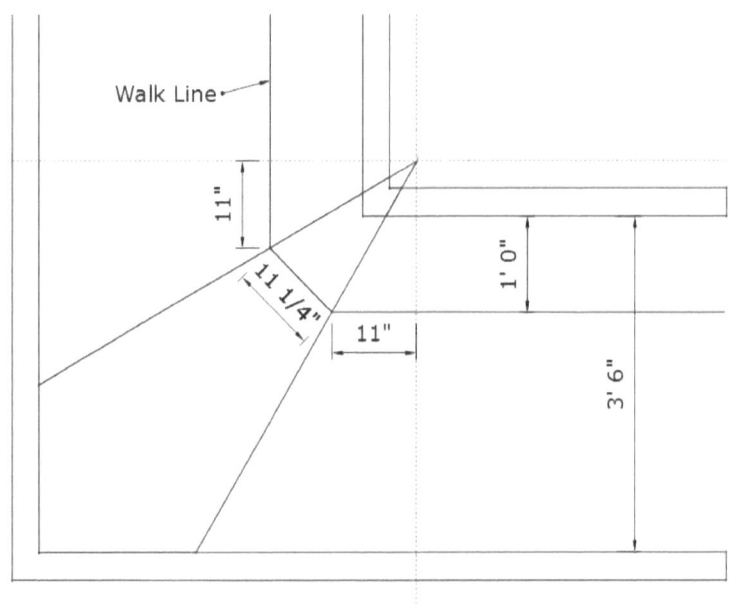

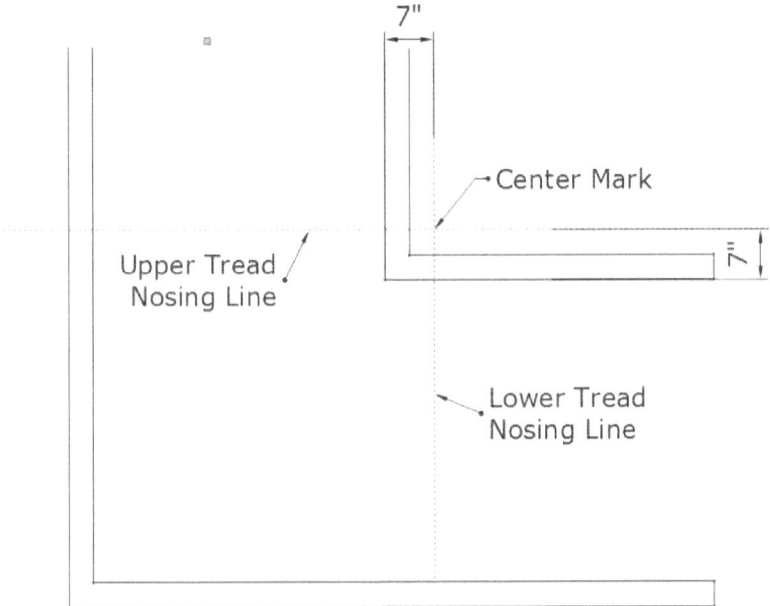

I'm going to provide you with two ways to figure out how to create equally proportioned angled steps. The first method will work for stairways with two or more winder treads.

The first thing will be to create a square with a center mark. This square can be created on a set of building plans or by the stair builder directly on the floor where the stairs will be built.

We are going to locate the center mark 7 inches from the inside of the stairway as shown in illustration above. The measurement can be larger than 7 inches for this type of design with three steps or treads, but I don't recommend using smaller measurements for safety reasons.

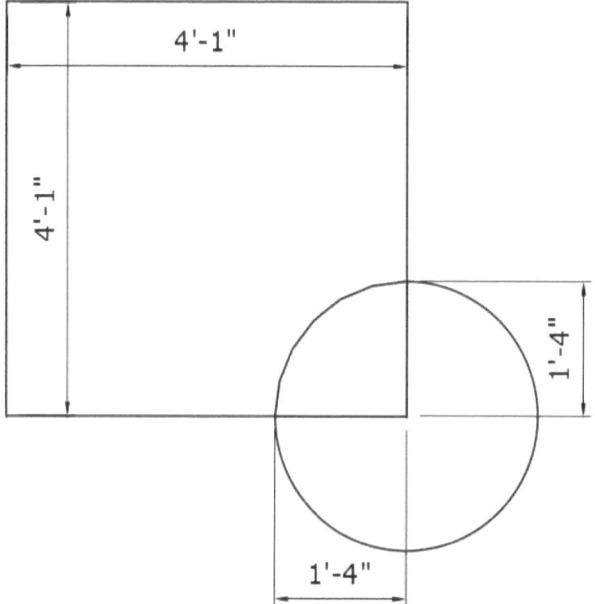

The next step will be to draw a circle inside your square using the center mark. The circle can be any size and after it's drawn divide ¼ of the circumference into three equal spaces.

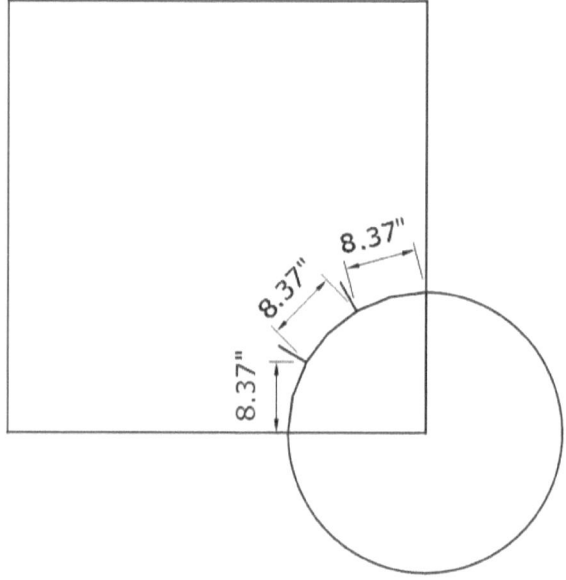

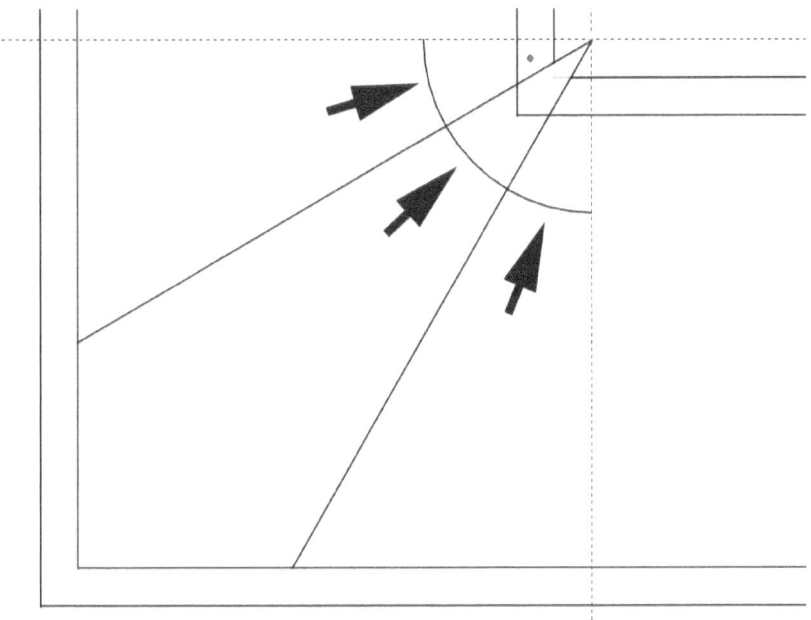

The two measurements dividing three equal spaces can now be used to draw the front of your tread or step lines. Line a straight edge up with the center mark to draw lines. To find the circumference of a circle or the outside line, multiply 3.14 times the diameter.

Example Using Previous Illustrations:

1'4" = 16" or the radius, which is half the diameter.

32 × 3.14 = 100.45

The circumference is 100.45 inches, but we're only going to be using one quarter of the circle.

100.45 ÷ 4 = 25.12

Divide this number by the amount of treads.

25.12 ÷ 3 equals 8.37 inches or 8 3/8 inches.

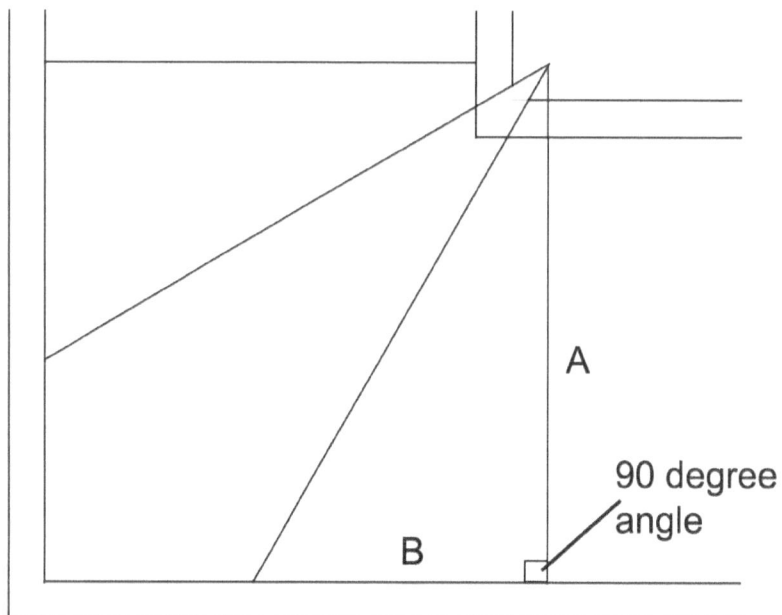

The second method requires one 90° angle and can be used to figure the first and third steps but not the middle of a three tread winder type stairway.

Side A Times 2
Divided By 1.73
Divided By 2 = Side B

Side A & B must intersect at 90° as shown in illustration above.

Example if Side A is 4 foot 1 inch or 49 inches then all you would need to do is multiply 49 times 2 equaling (98) then dividing 98 by 1.73 equaling (56.65) then divide 56.65 by 2 equaling (28.32).

The final answer for Side B is 28-3/8" or 28.32 inches.

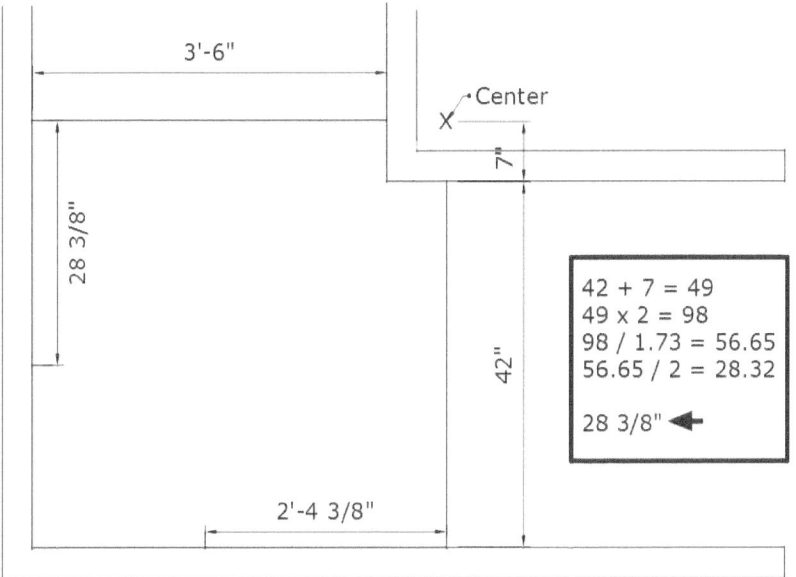

Line up side B measurement with center mark to draw front of tread line for winders The illustration above uses a 7 inch center mark and the one below uses a 10 inch center mark.

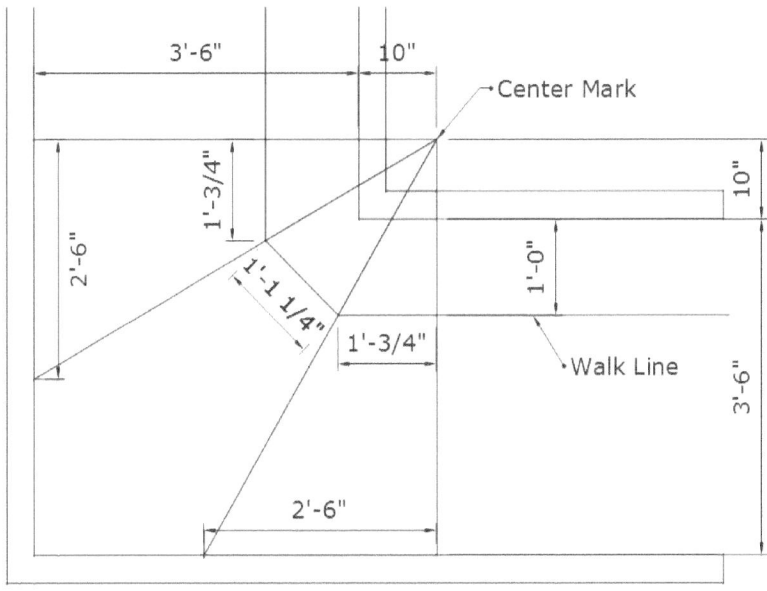

Winder Examples

Example 1: Using Supporting Walls

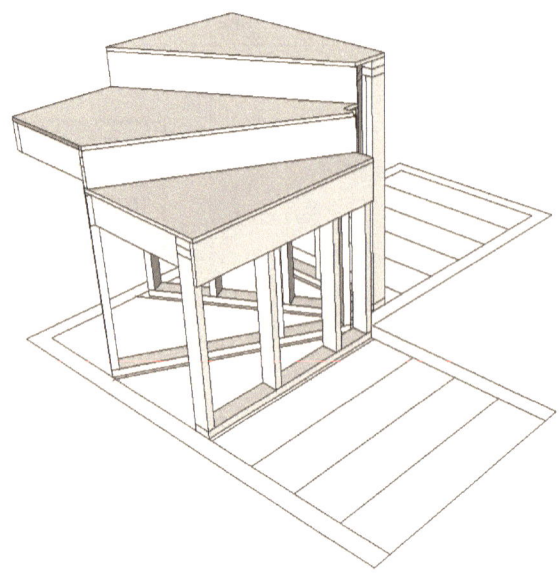

In this example we're going to use supporting walls to hold up the winder steps and stringers.

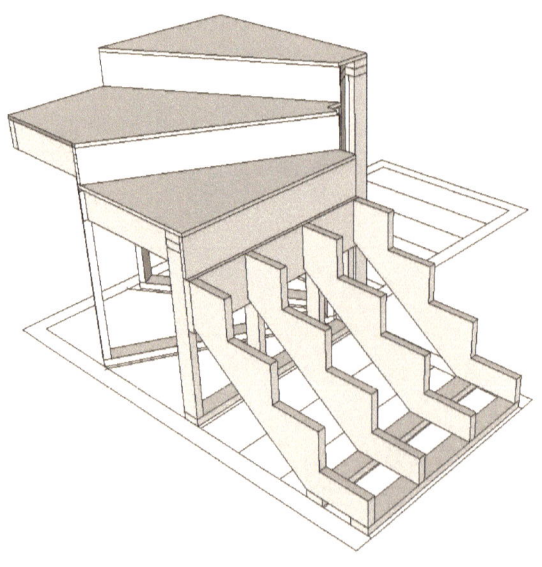

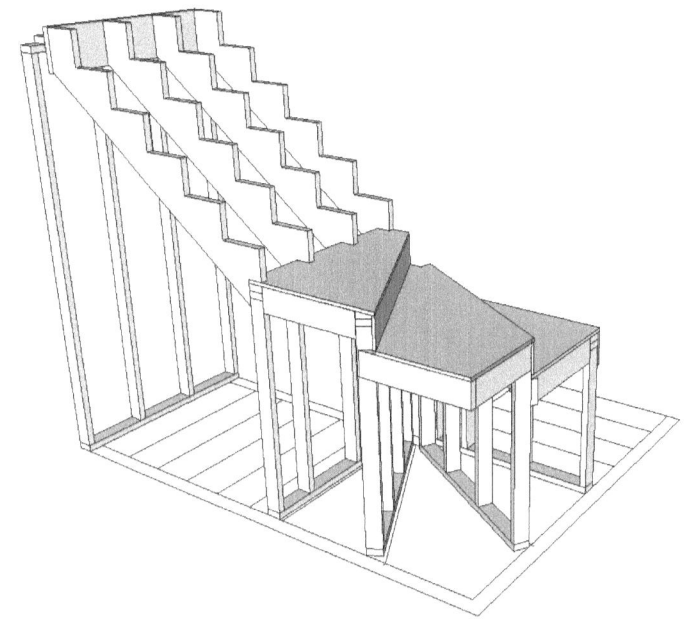

Each supporting wall will be moved back ¾" (thickness of riser) from the line on the foundation that represents the face of step.

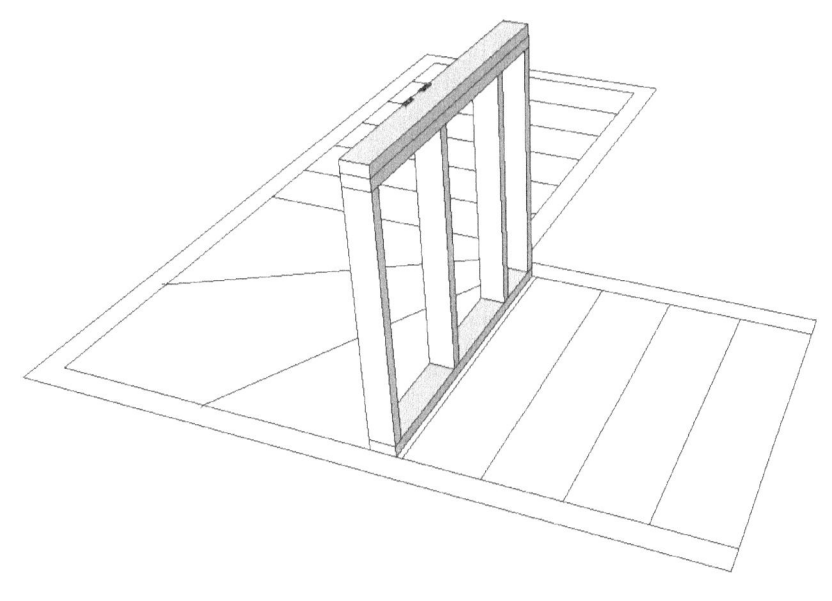

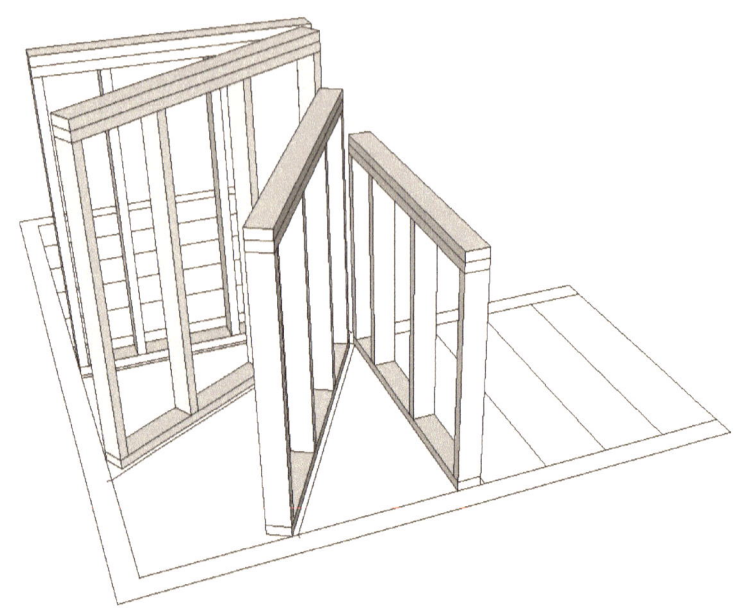

The first three walls represent one step and the fourth wall which is level with the third wall will support a notched stringer and landing joist.

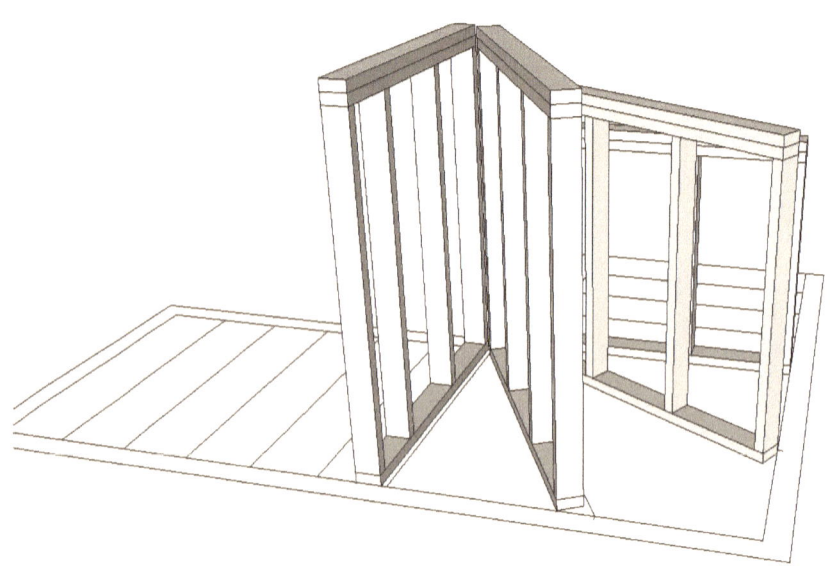

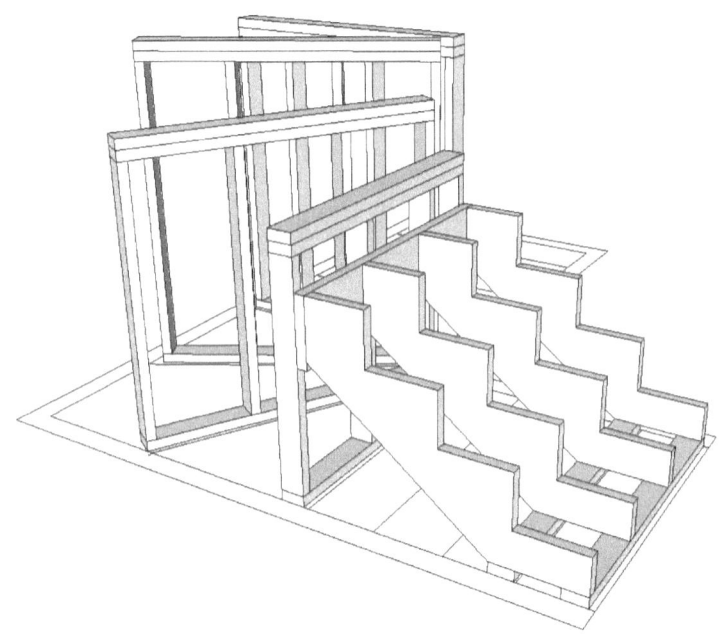

These illustrations are using a ledger to attach stringers to winder supporting wall and baseplates to connect them to the building foundation.

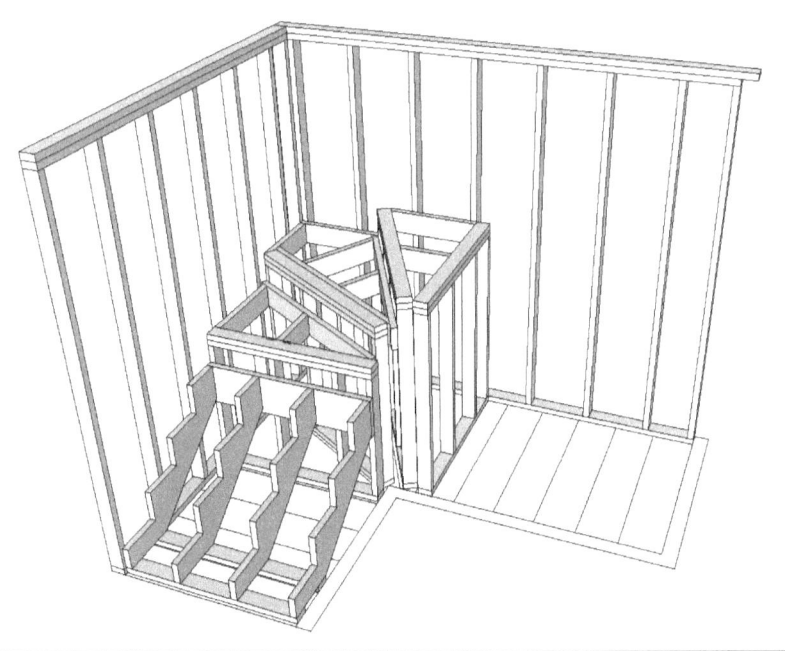

These drawings provide you with two different ways to frame winder steps. The lower step uses a joist hanger to attach to the wall and the middle uses a ledger without hangers.

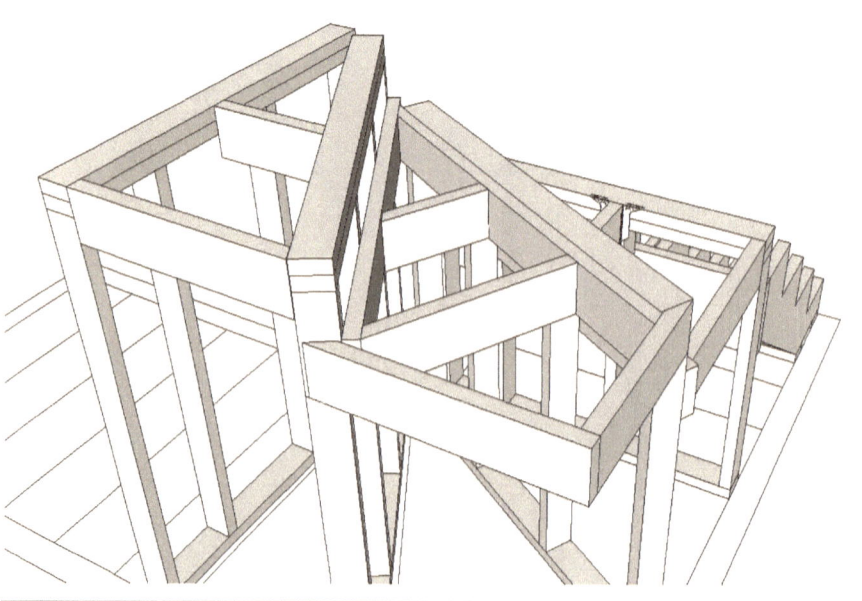

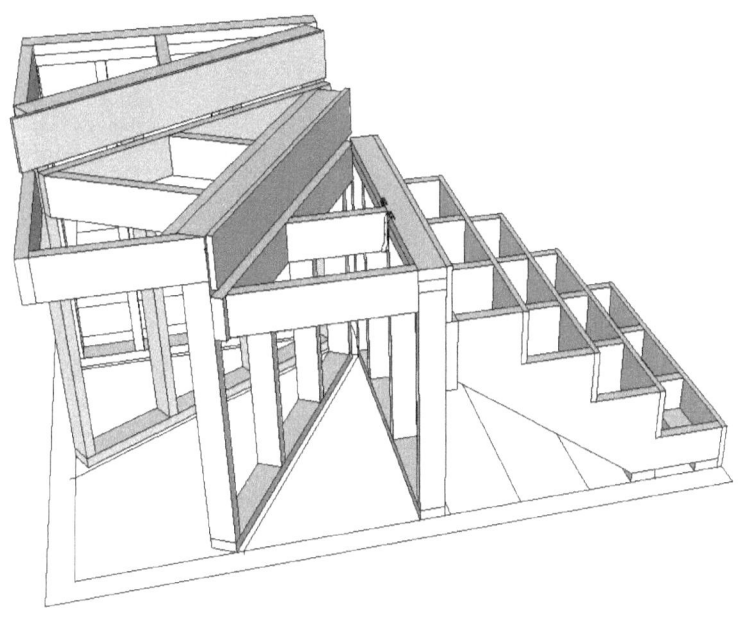

The illustration above shows the ¾" risers and the illustration below shows the ¾" plywood treads or steps.

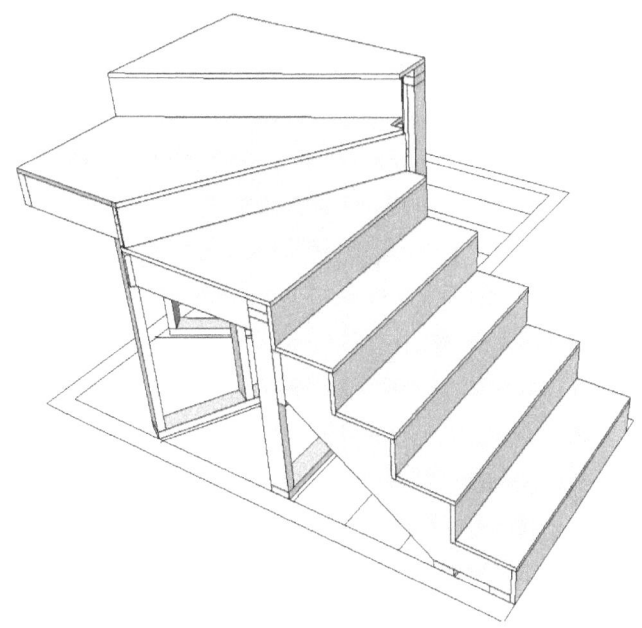

The upper stringers will sit on top of the last winder step and attach to the upper floor supporting wall with a ledger.

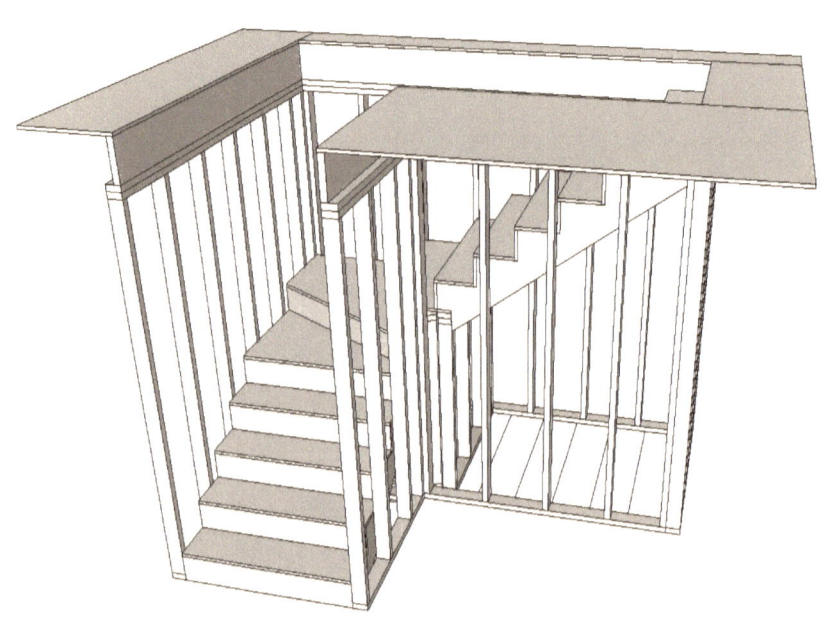

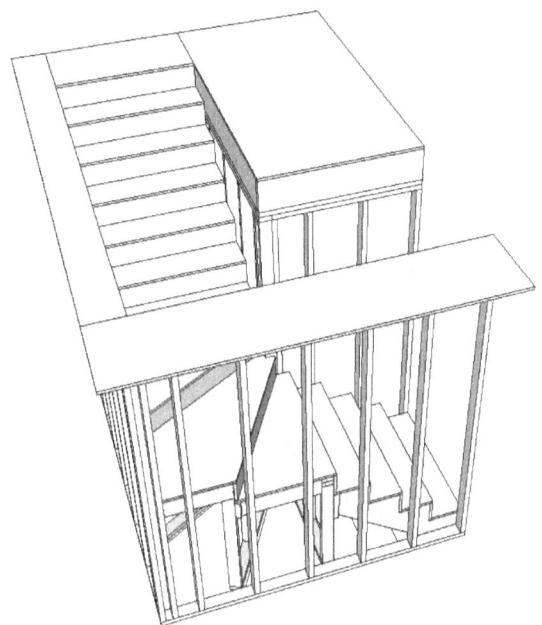

These illustrations include the interior walls that were removed to provide you with a better view. Parts of the stairway including the winder step joist can also attach to these walls.

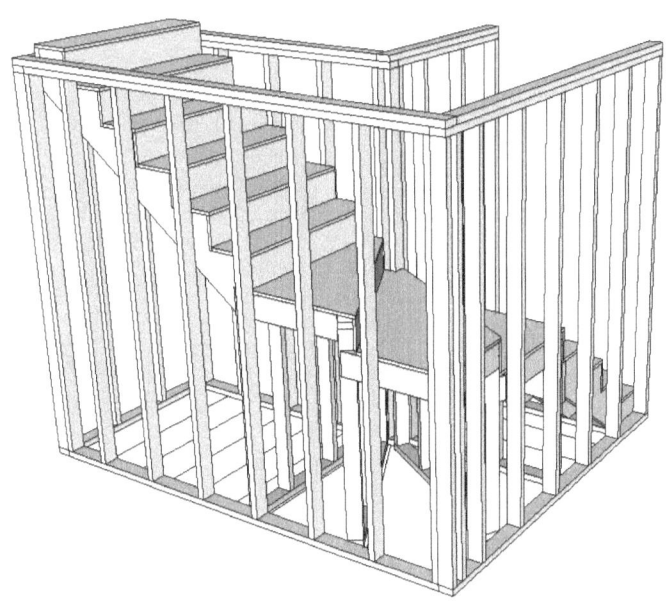

Example 2: Lower and Upper Winders

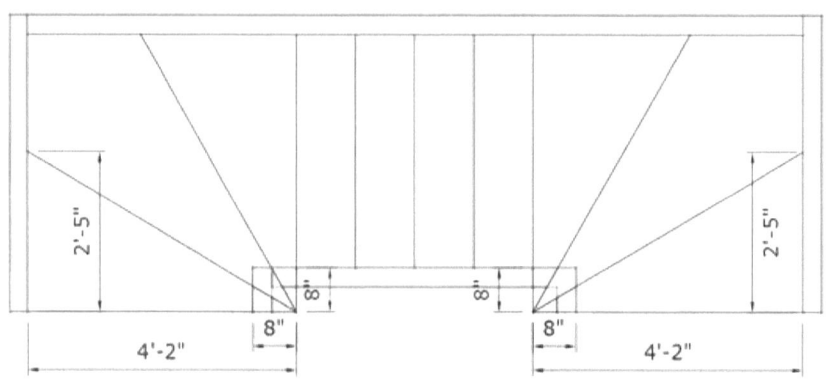

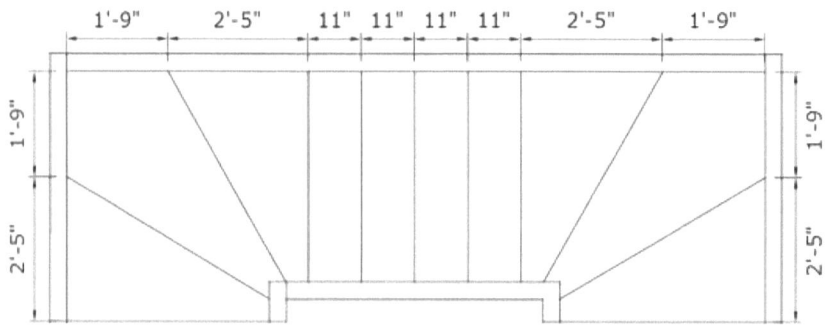

Three different floor plan illustrations with measurements.

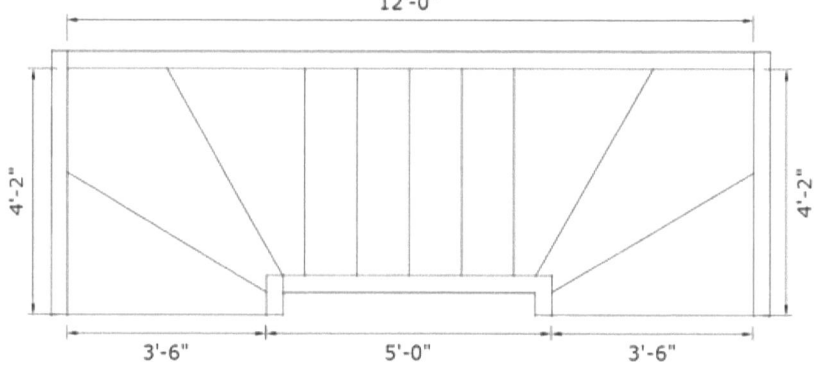

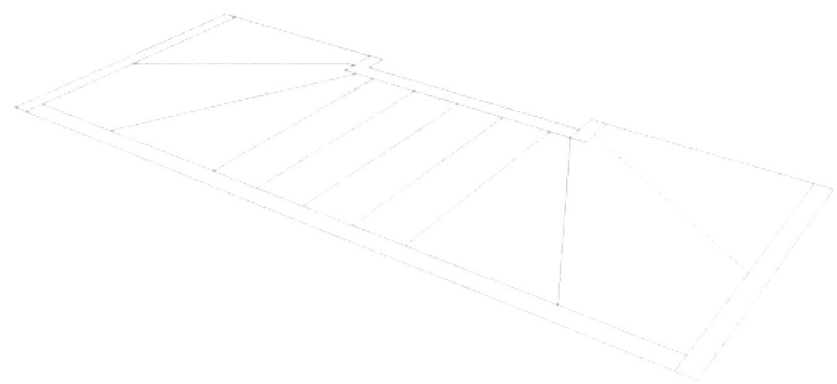

In this example we're going to set each winder step on top of the one beneath and use stringers to connect the lower and upper winder steps.

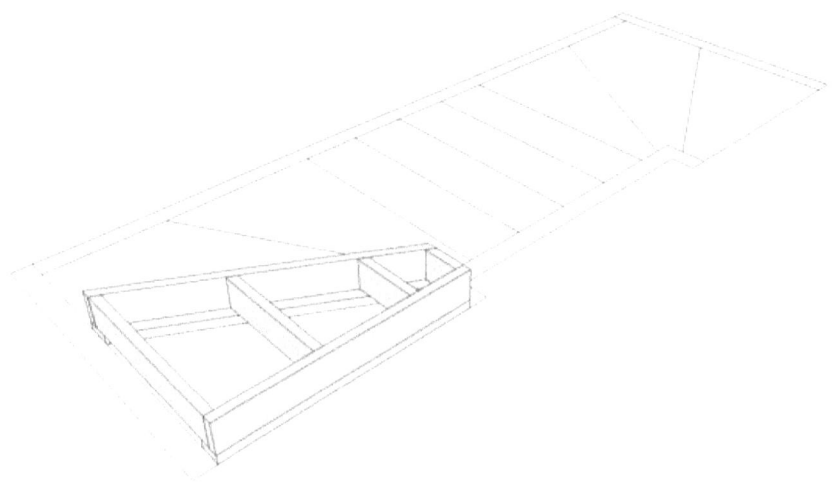

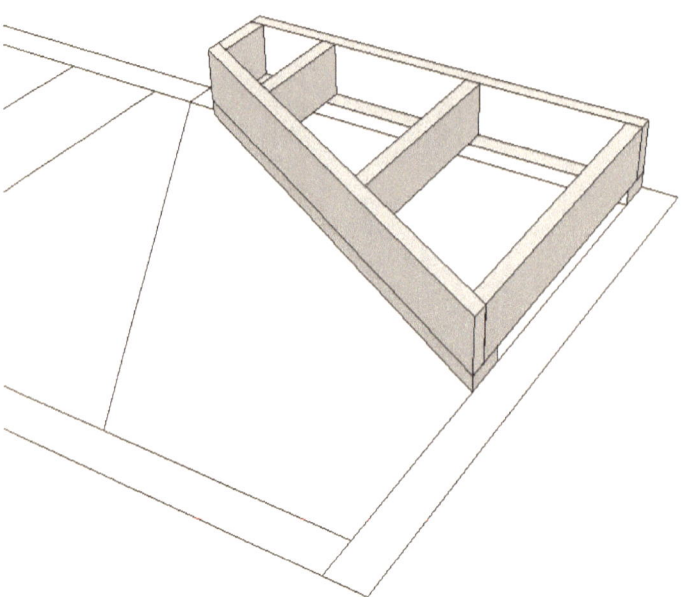

This method might require you to resize the winder joist. For example a 7 ½ inch step with ¾" plywood sheathing sitting on top of a 1 ½" baseplate will need to be resized to 5 ¼ inches.

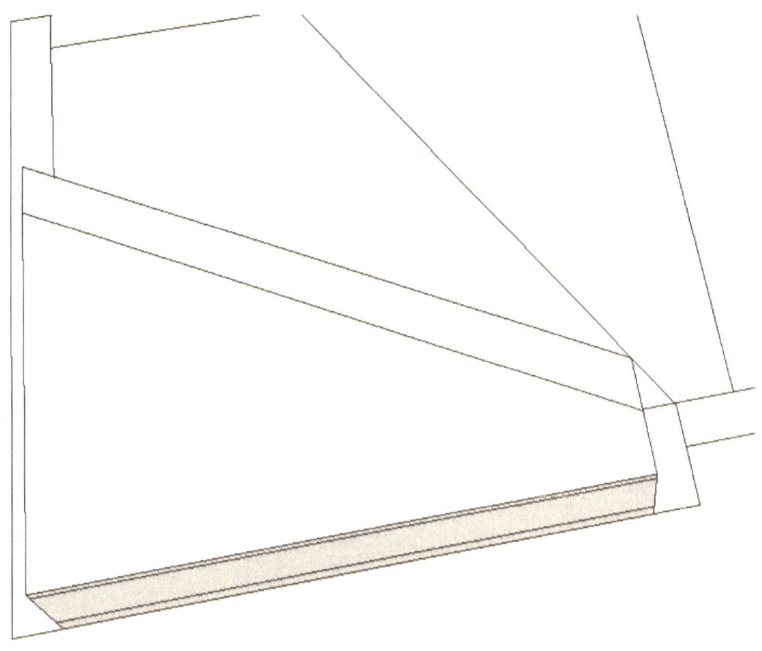

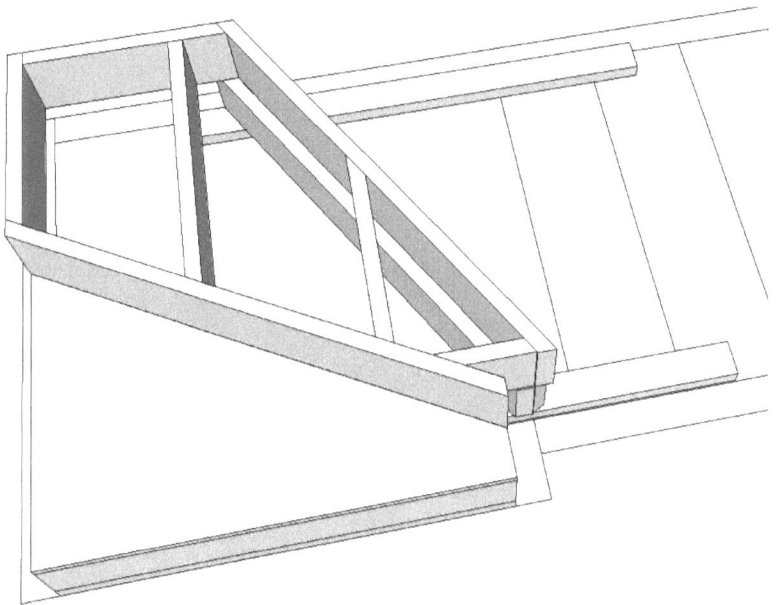

The second step will sit on top of the first step with a double joist support in the back. This design won't require a separate plywood riser, instead will be using the joist as the face of the first, second and third winder step.

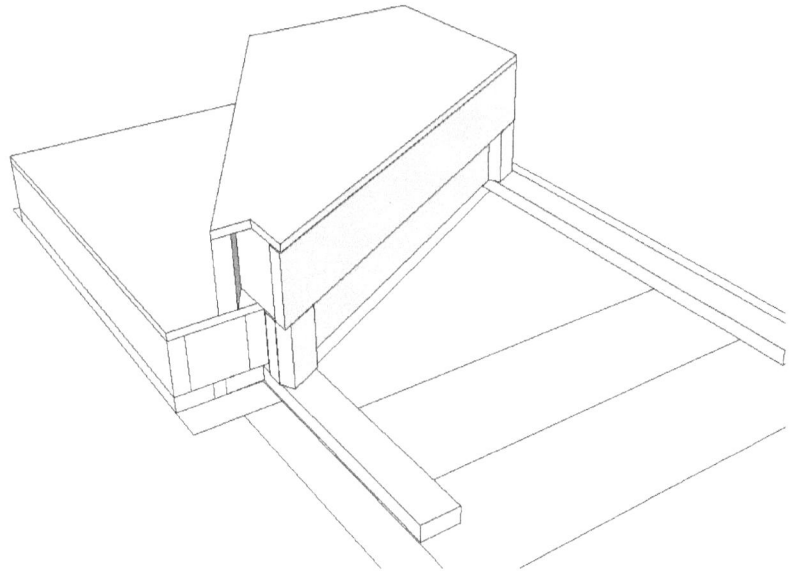

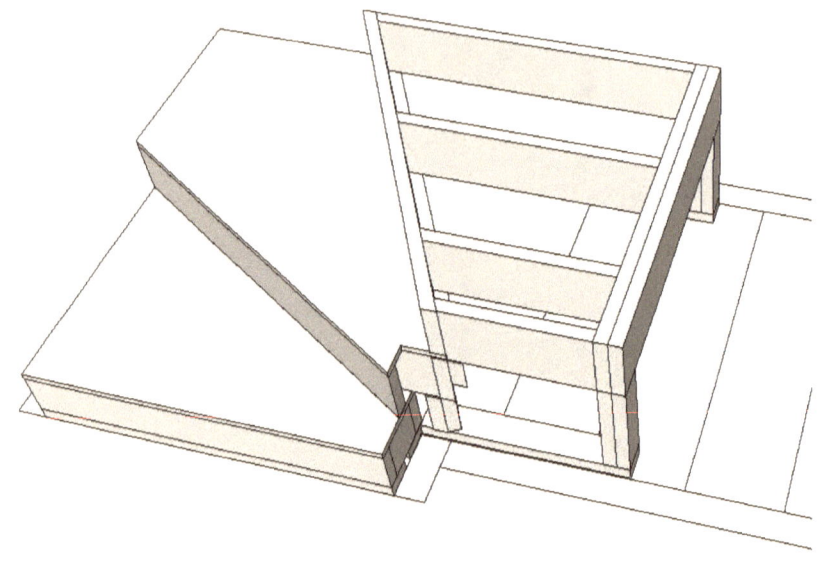

The third step is going to be built differently. The back part of the step will have double joist sitting on top of double 2 x 4 supporting posts. The front part of the third step will be supported by the second step along with a ledger to support the joist.

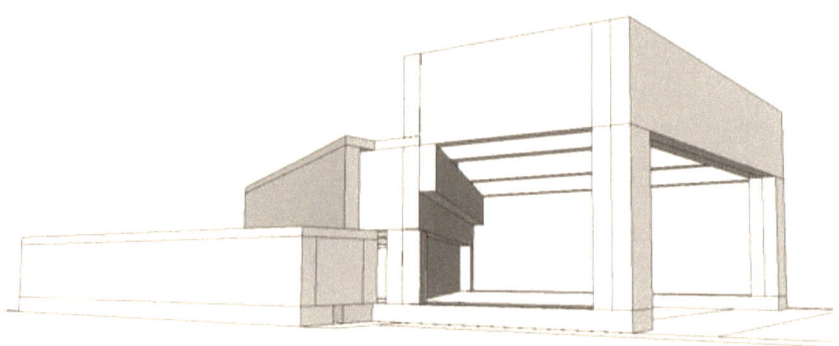

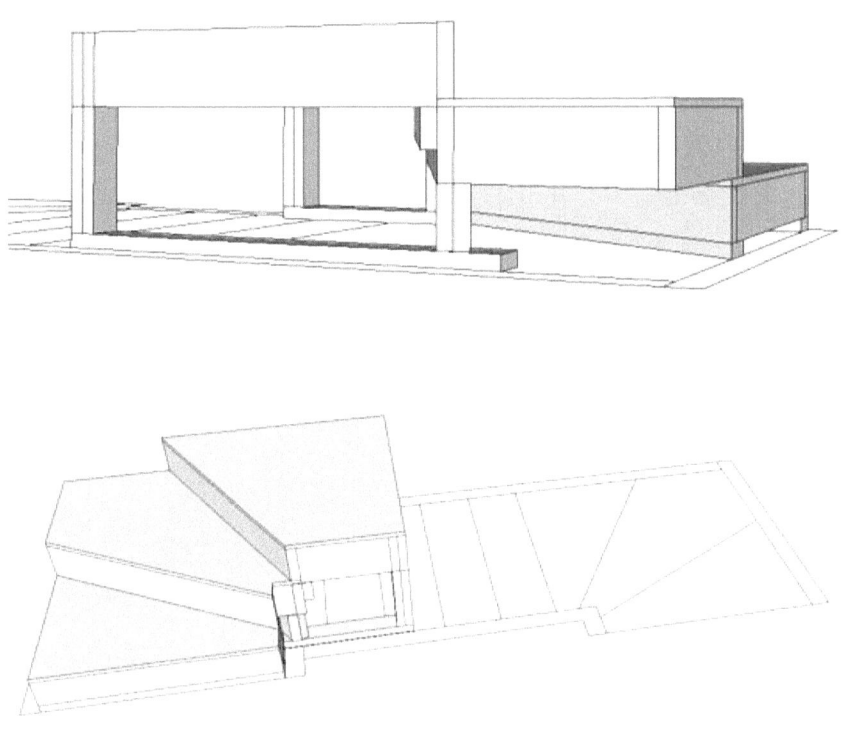

When completed the first three steps should look something like this.

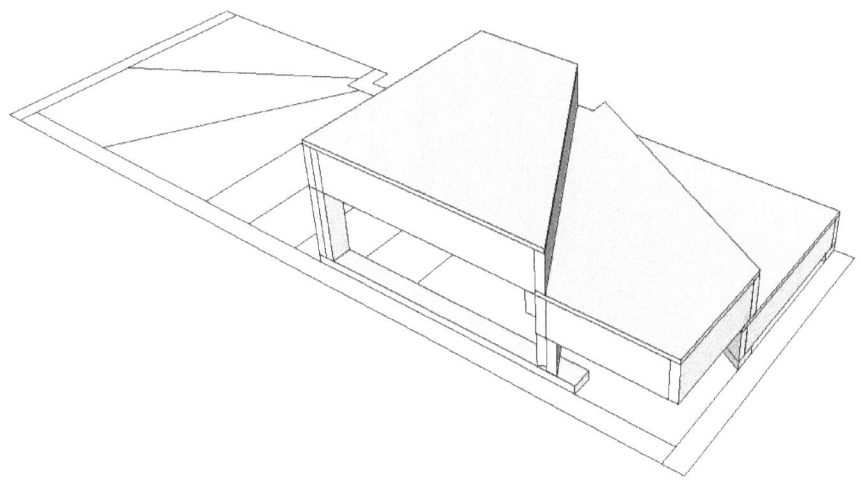

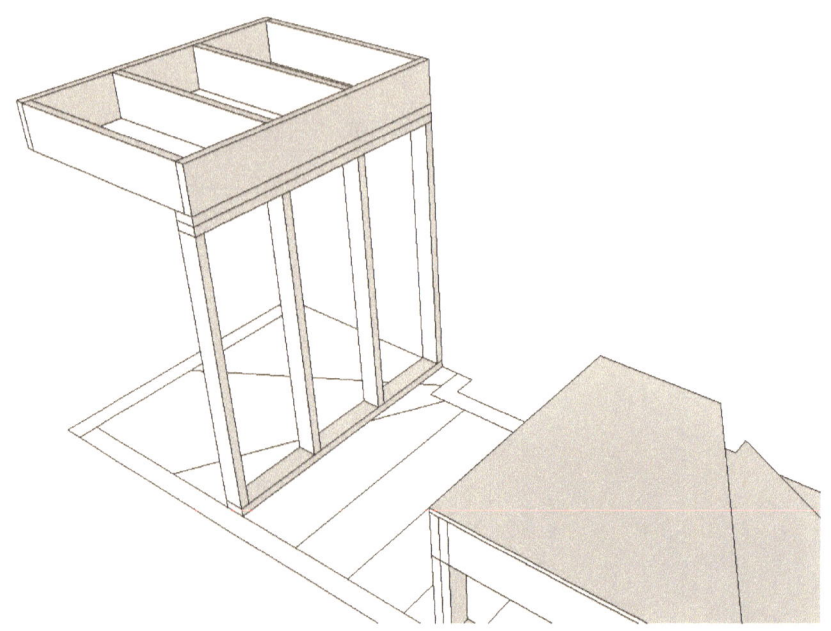

The upper landing will sit on top of a supporting wall and the other side will attach to upper level supporting walls that will be shown at the end of this example.

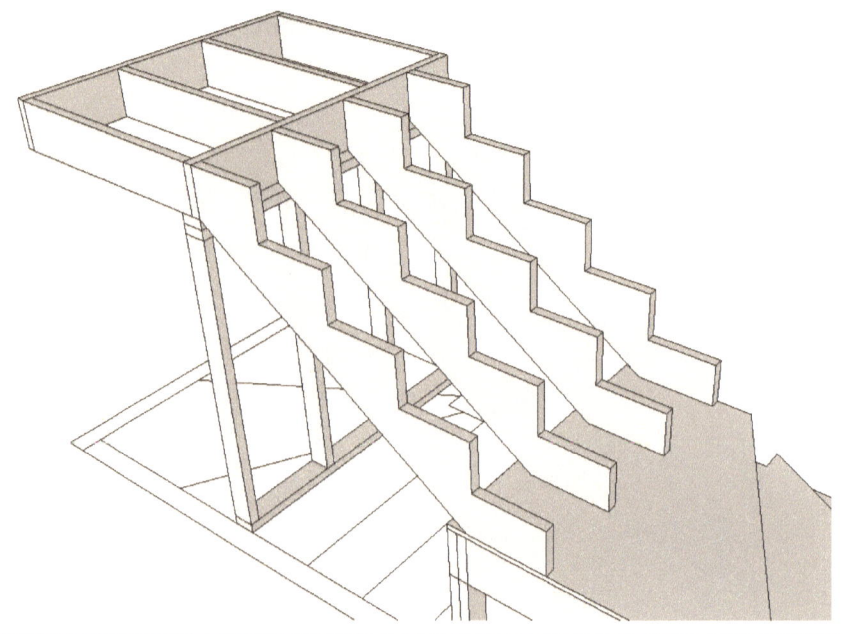

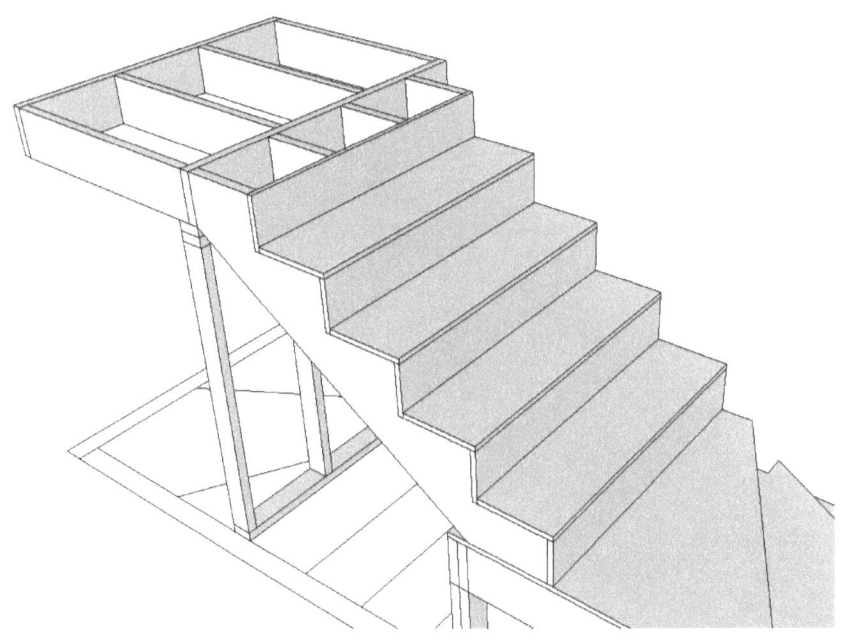

The stringers will need to be moved back horizontally to allow for the ¾" thickness of plywood risers.

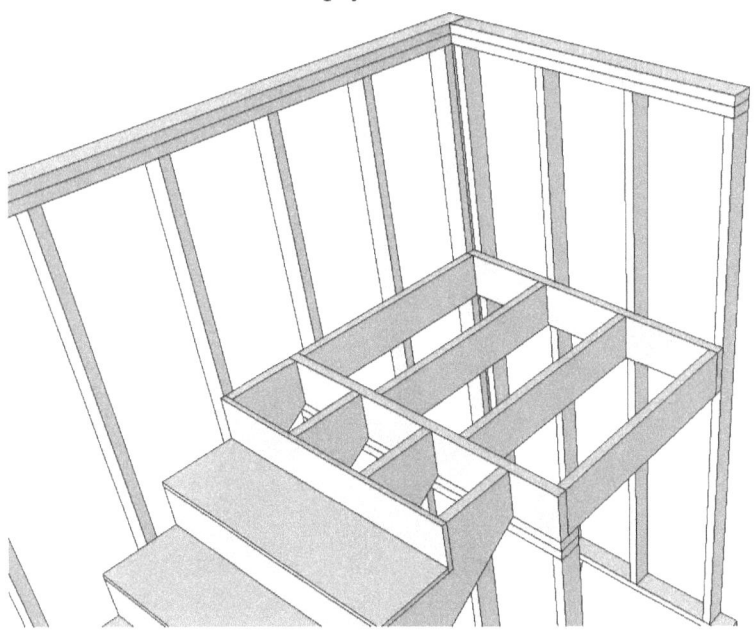

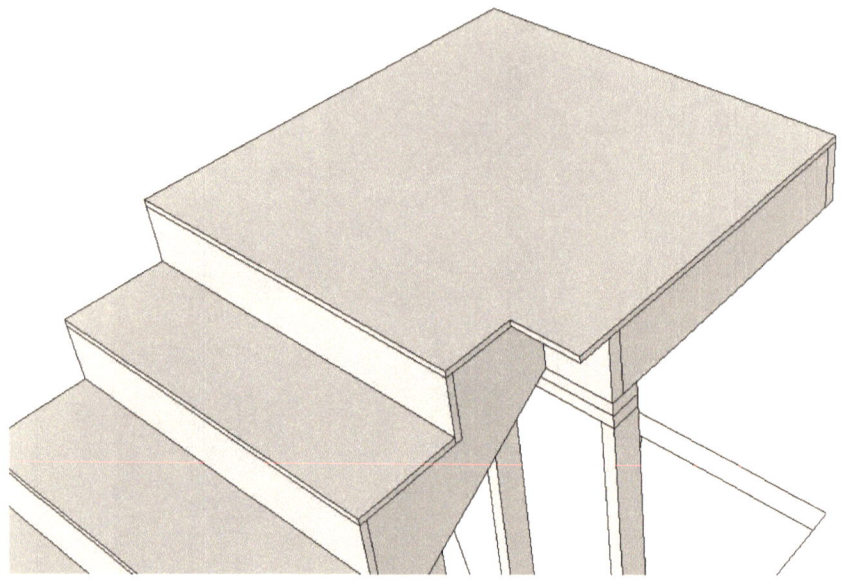

Install plywood sheathing and build next winder step. You can transfer face of step measurements to the landing, by measuring from the inside corner.

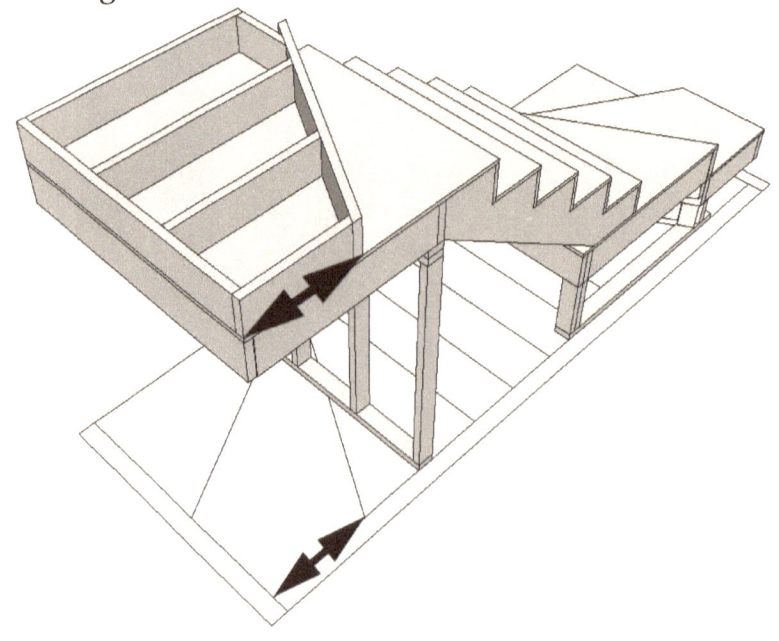

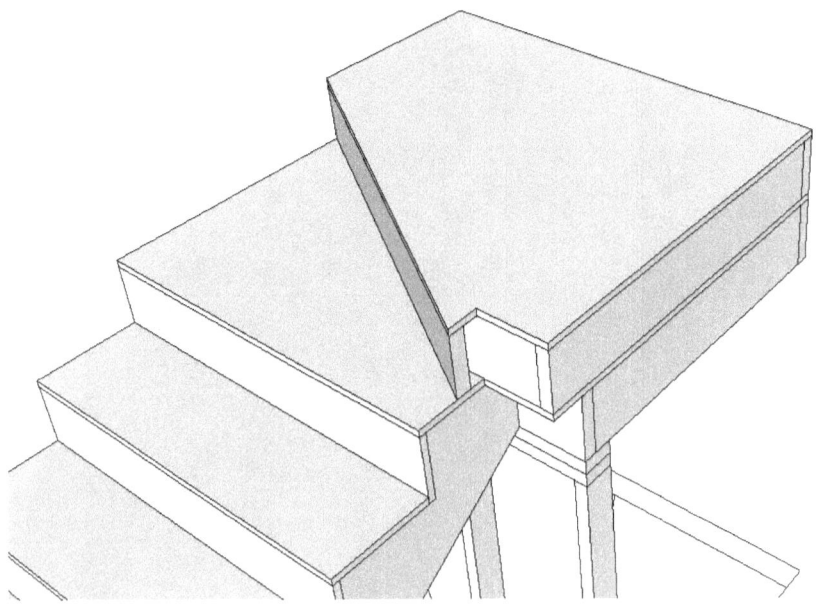

Install sheathing and build last step.

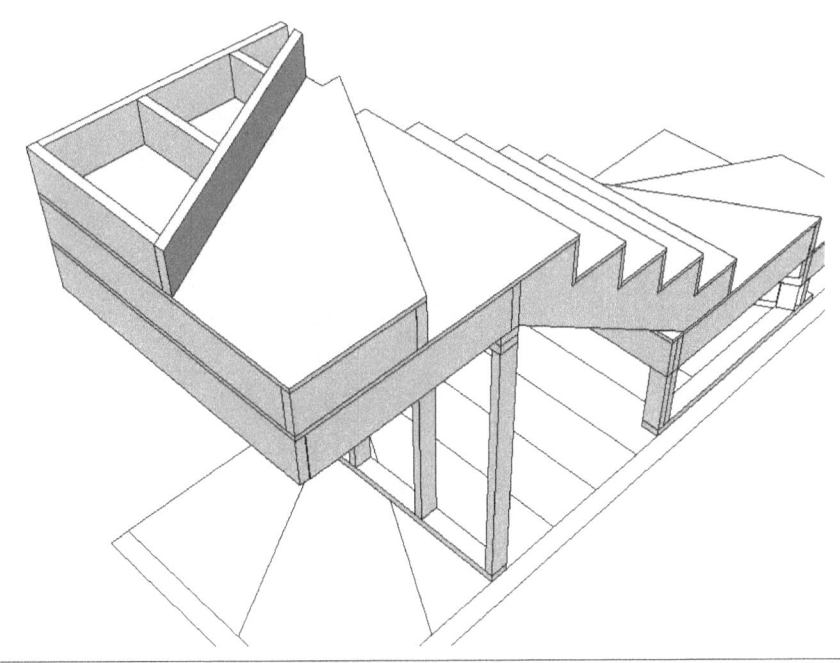

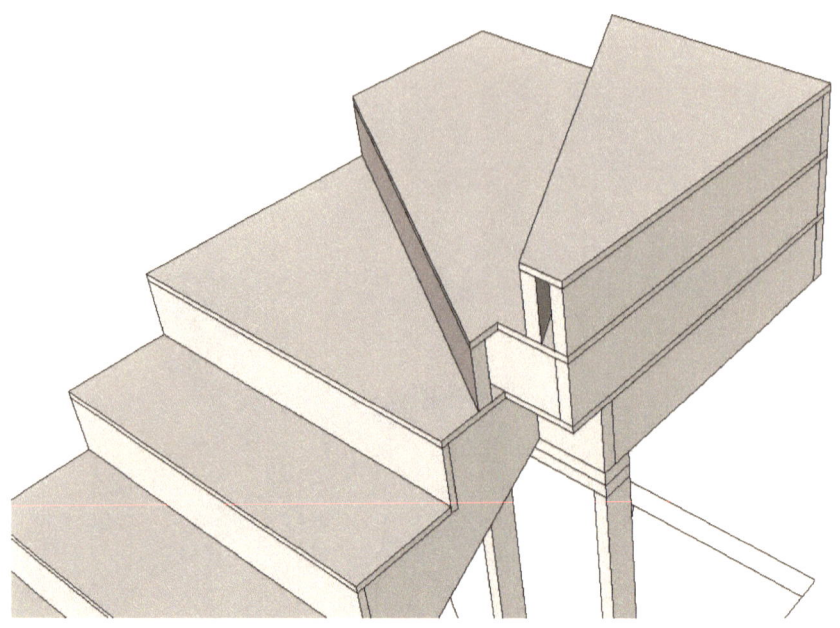

In this example I provided you with two different ways to build winder steps and the upper framing method can be used at the bottom or middle of the stairway also.

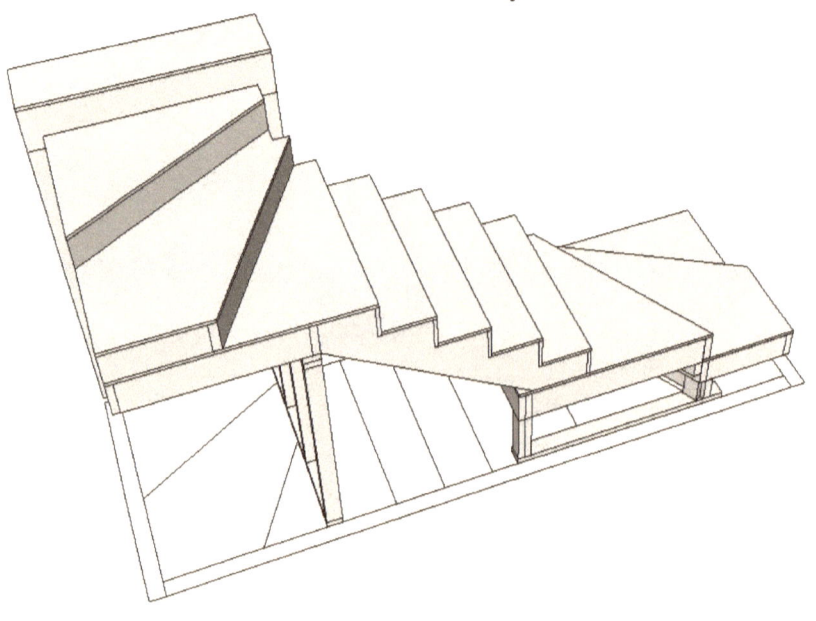

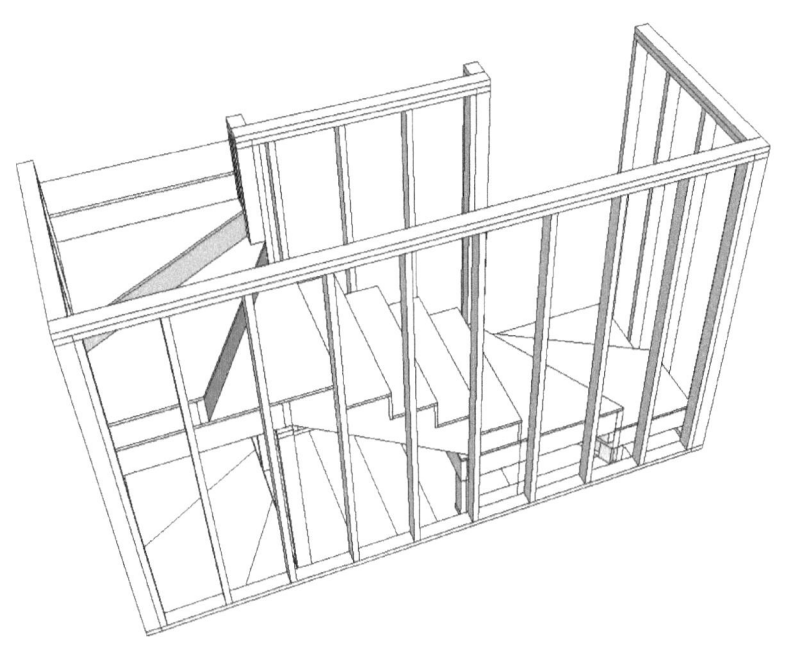

These illustrations provide us with the upper level supporting walls. Remember, winder joist can usually be attached to other walls.

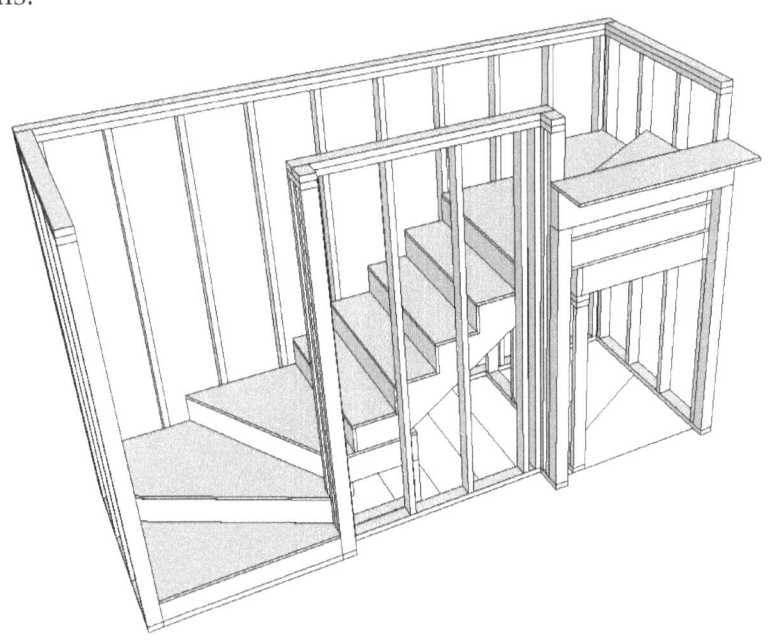

Example 3: Five Step Winders

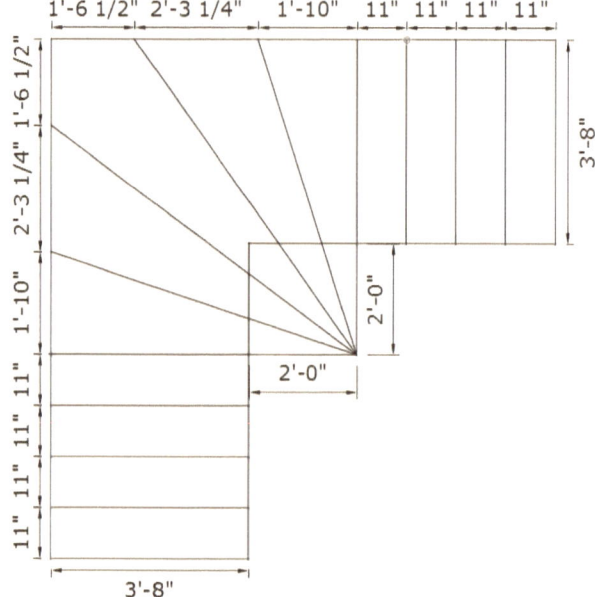

Illustration above is a simple floor plan and the illustration below shows the face of steps marked on base plate.

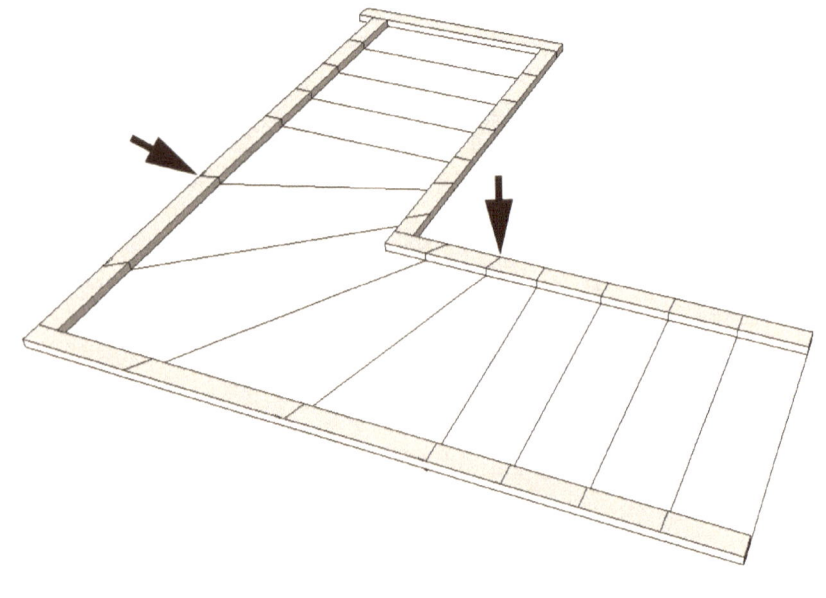

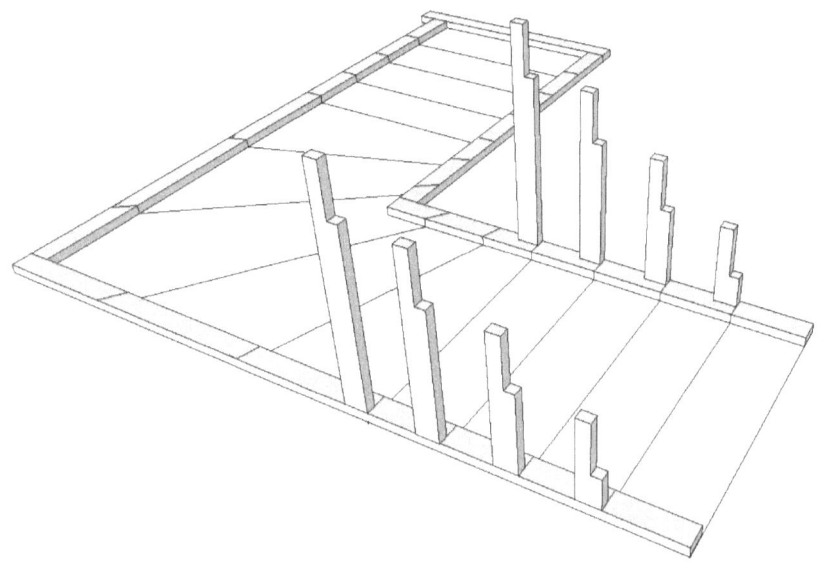

Notched 2 x 4 supports for 2 x 8 will line up with the face of step marks on base plates. Subtract tread (1 1/8") and base plate (1 ½") of each individual step to figure the height of notched 2 x 4 supports.

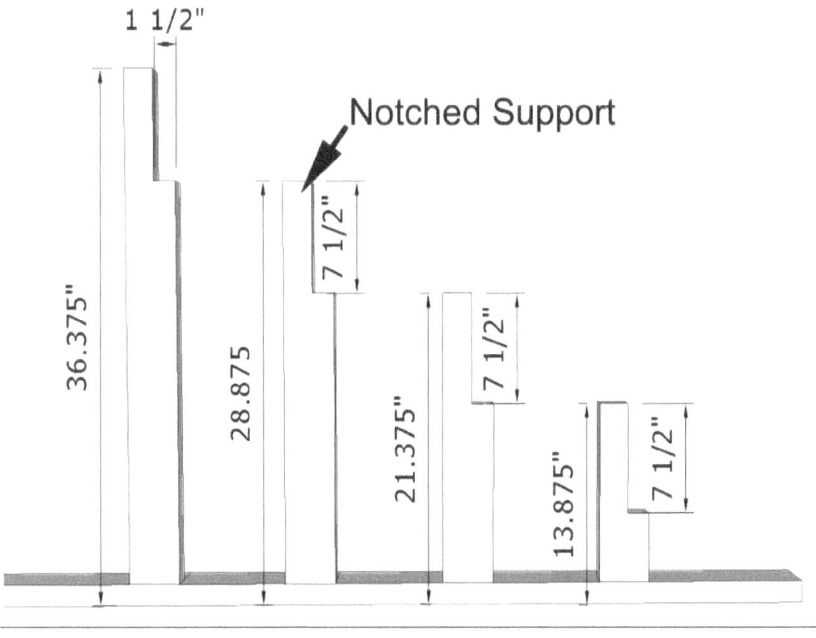

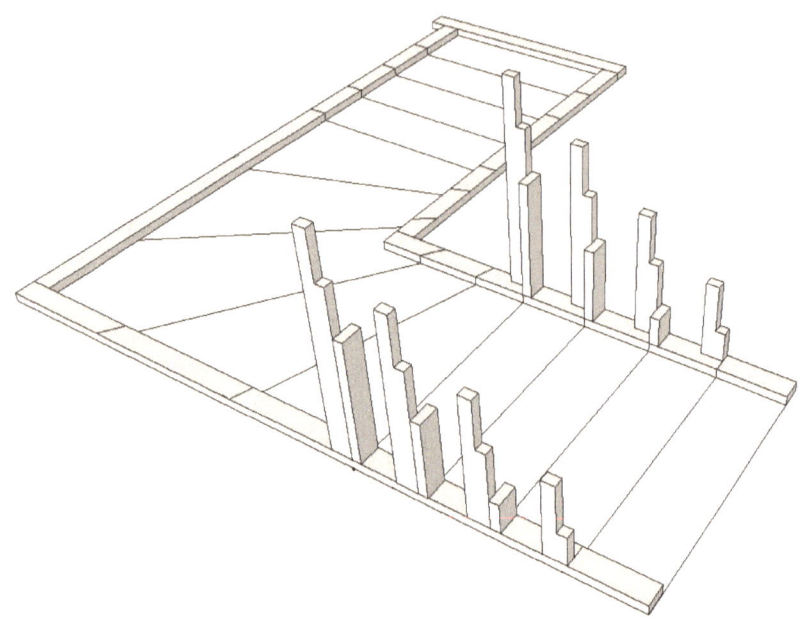

Subtract the height of step support framing (in our example we're using 2 x 8 which is 7 ½") from the height of front notched 2 x 4 supports to figure the height of 2 x 4 back of step supports.

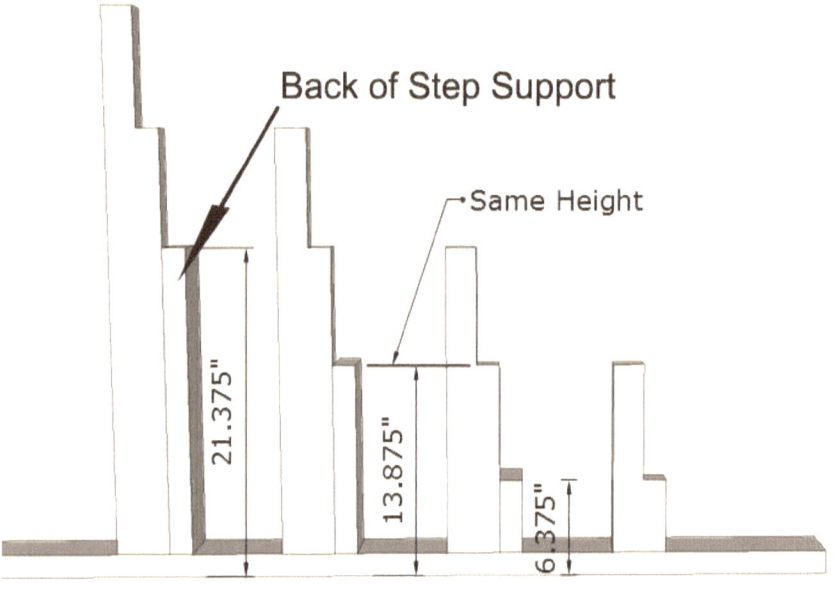

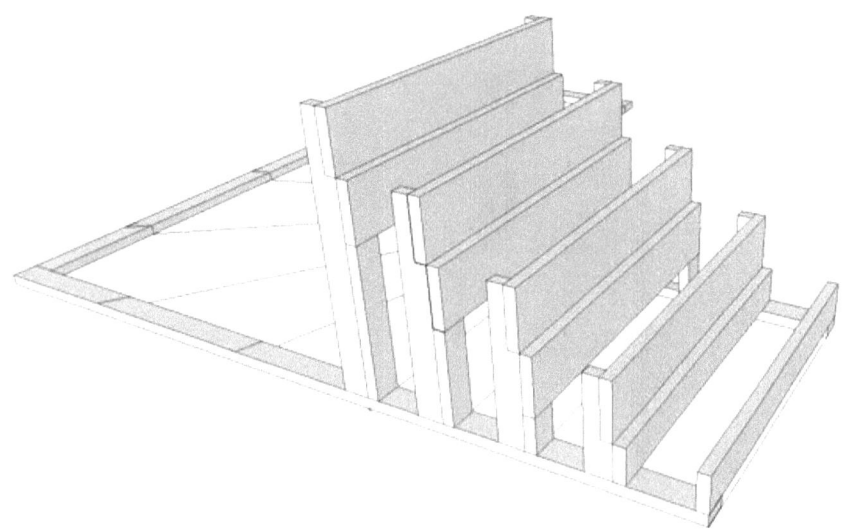

Our individual riser height for this example is 7 ½ inches. I would suggest building this type of stairway in sections. We've just completed the lower section.

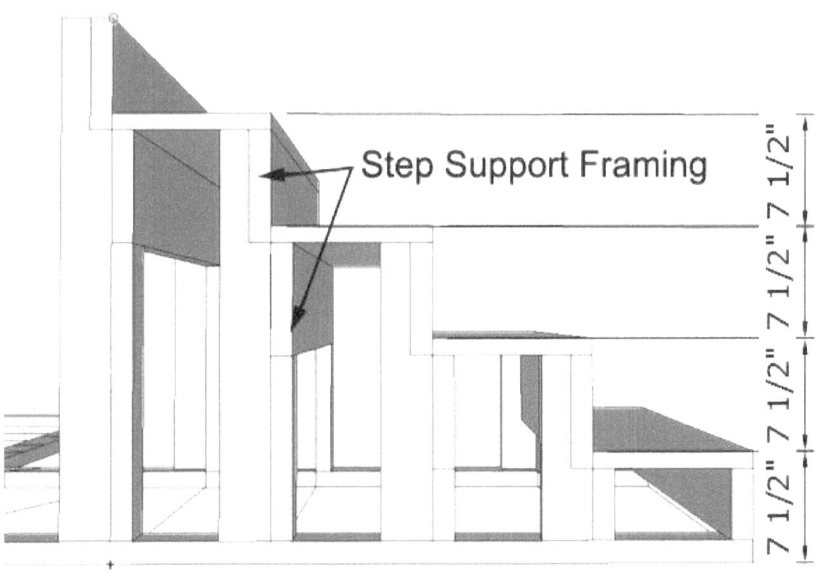

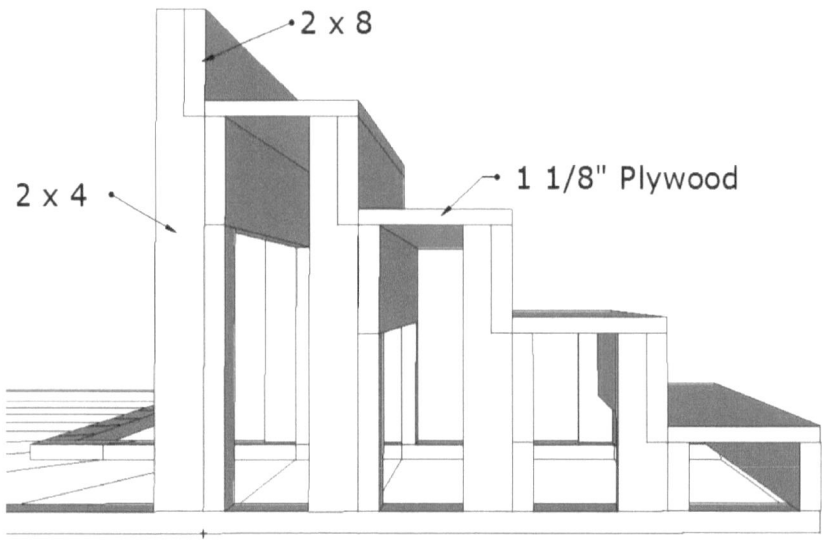

Double check all of your measurements at this point. You must be extremely careful when building this type of stairway to get all of your parts in the right spot. A common problem with stair builders using this method will be to install the back of step supports where the notched supports should be on the face of step line.

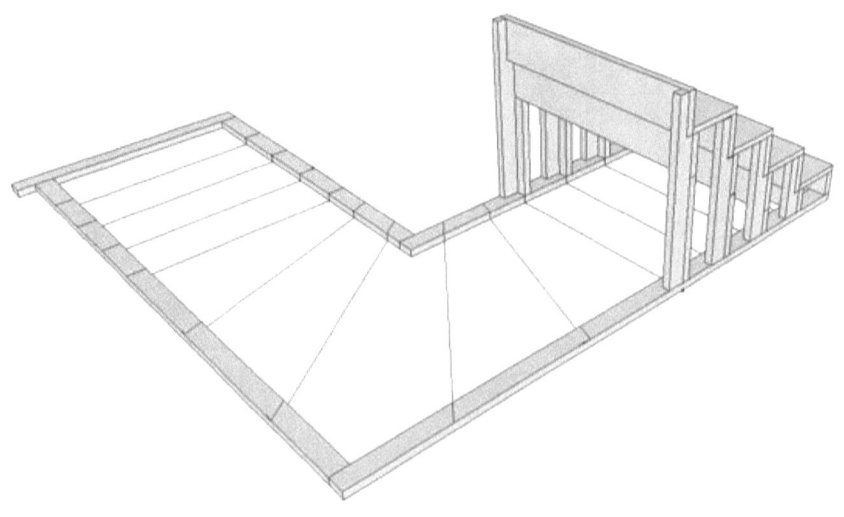

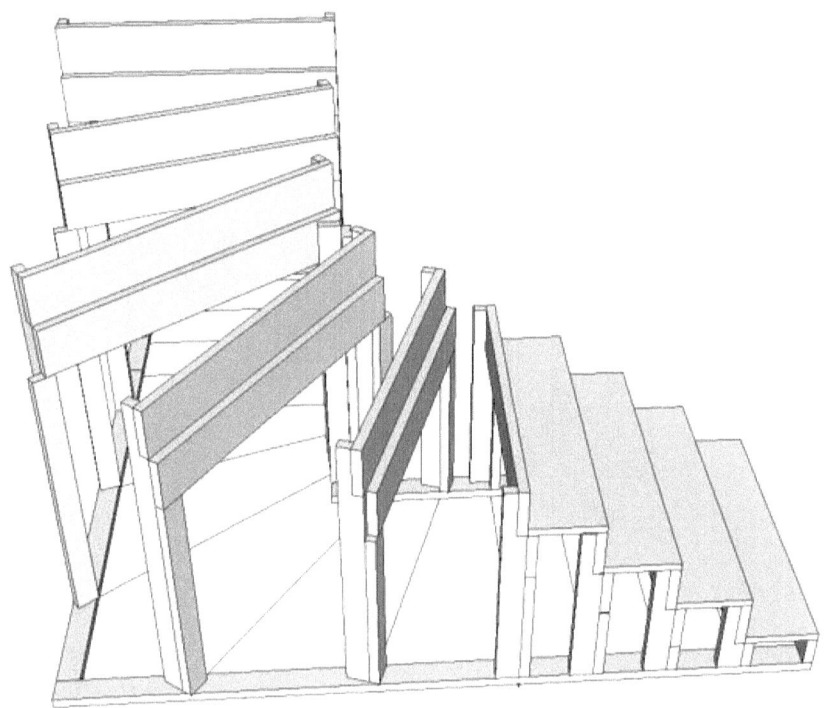

Use the same instructions to assemble the middle section as you did in the lower section.

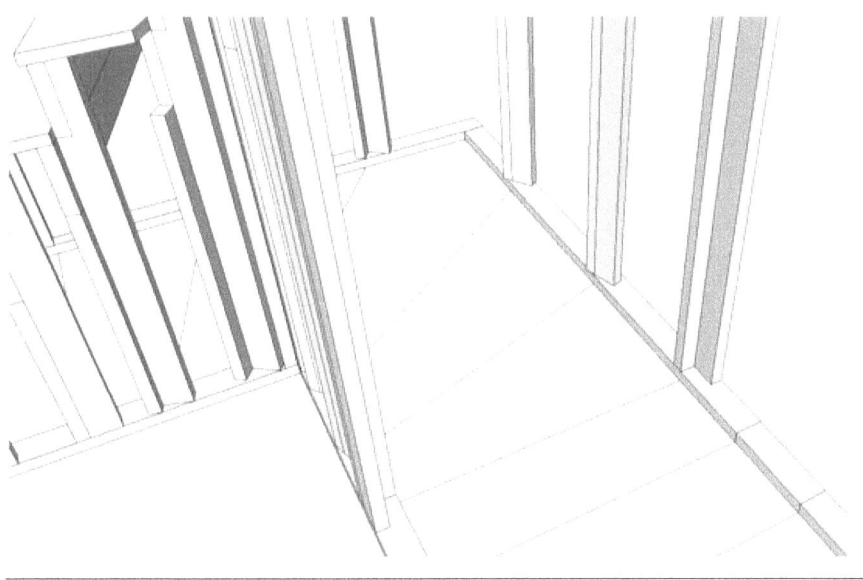

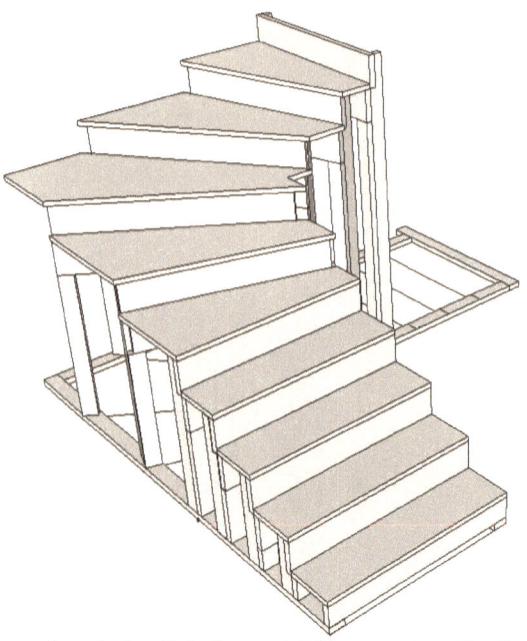

Make sure you don't build the middle section before laying out your stair treads. This information can be found in the back of the book.

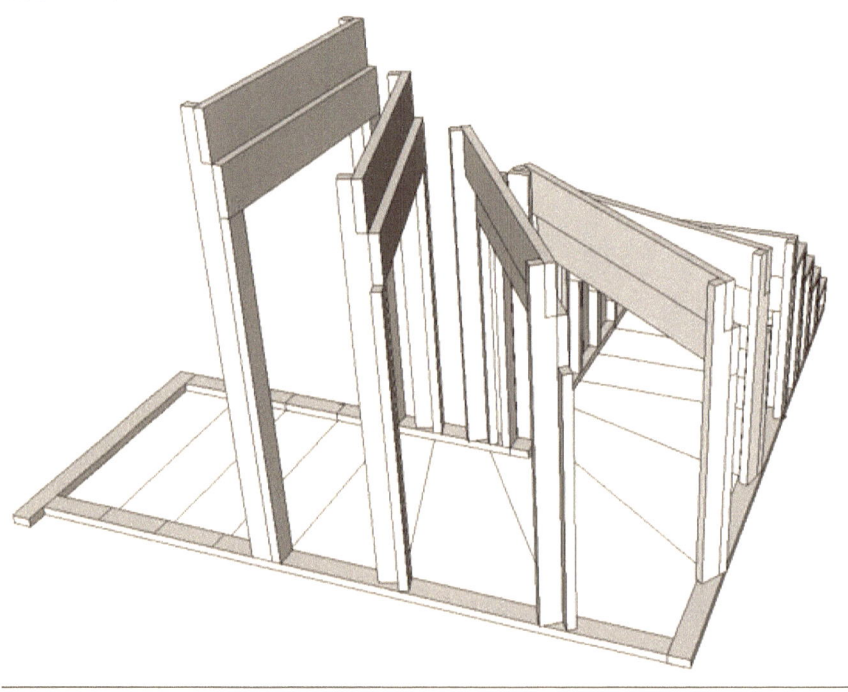

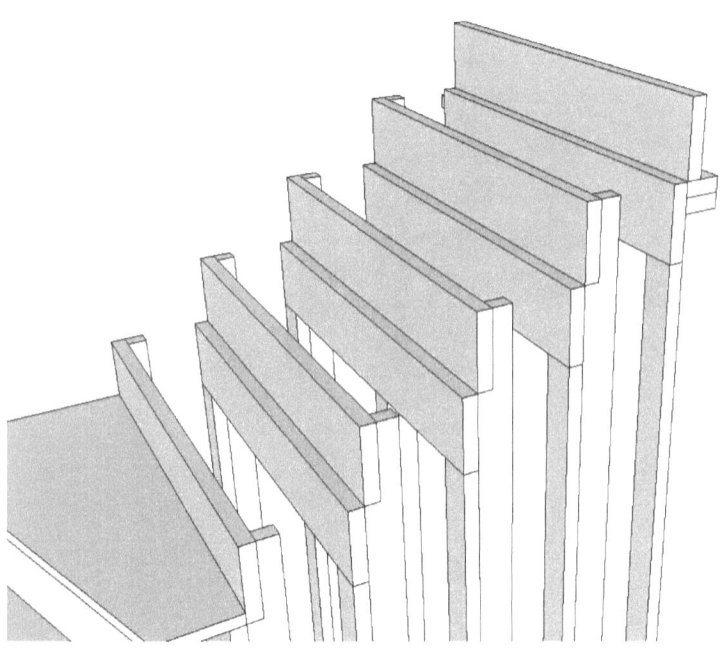

Use the same instructions to assemble the upper section as you did in the lower and middle sections.

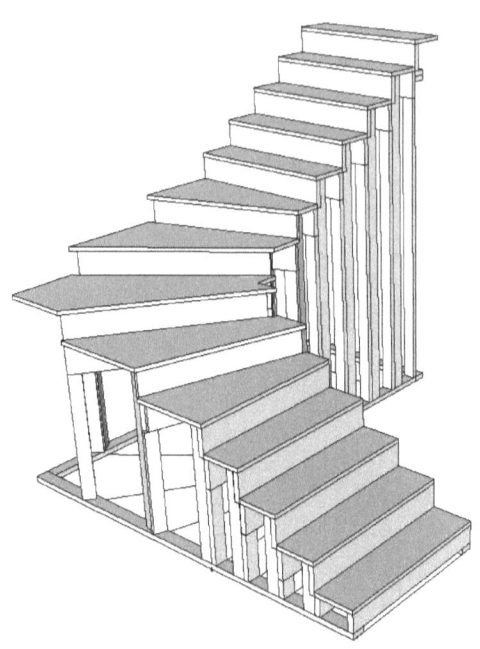

Stairway should look something like this when completed.

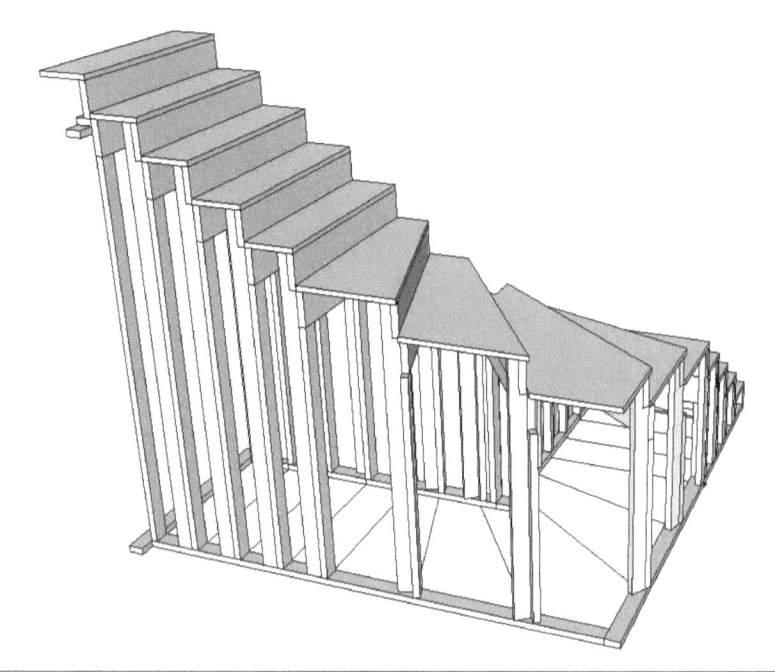

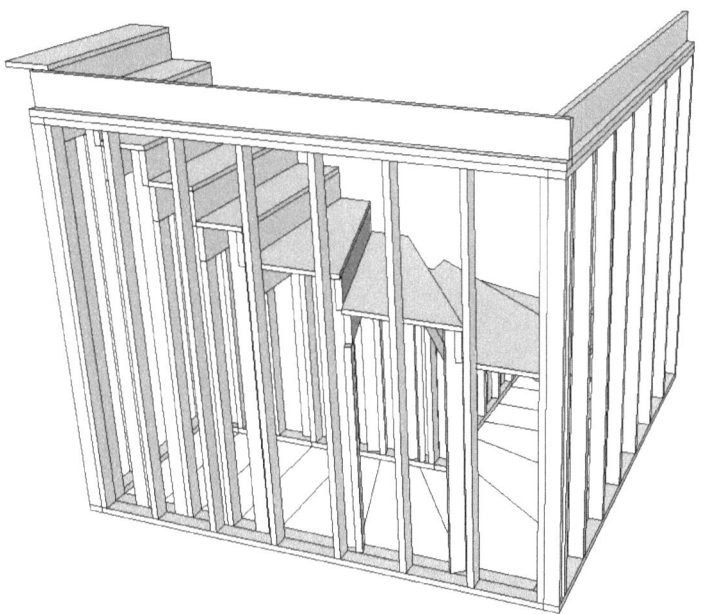

Here's the stairway with walls that will be used to support the upper floor. Remember, you can use these walls to attach parts of the stairway framing.

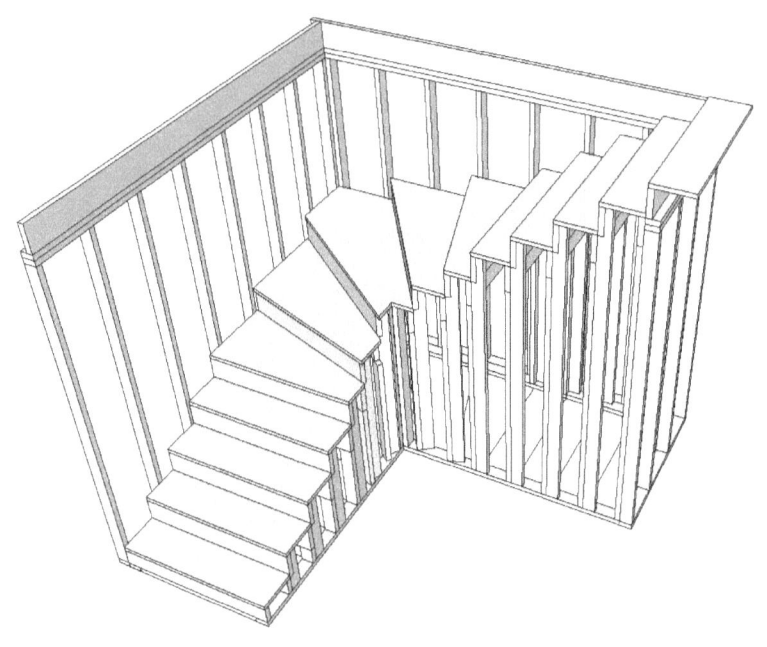

Additional Illustrations of Example 3

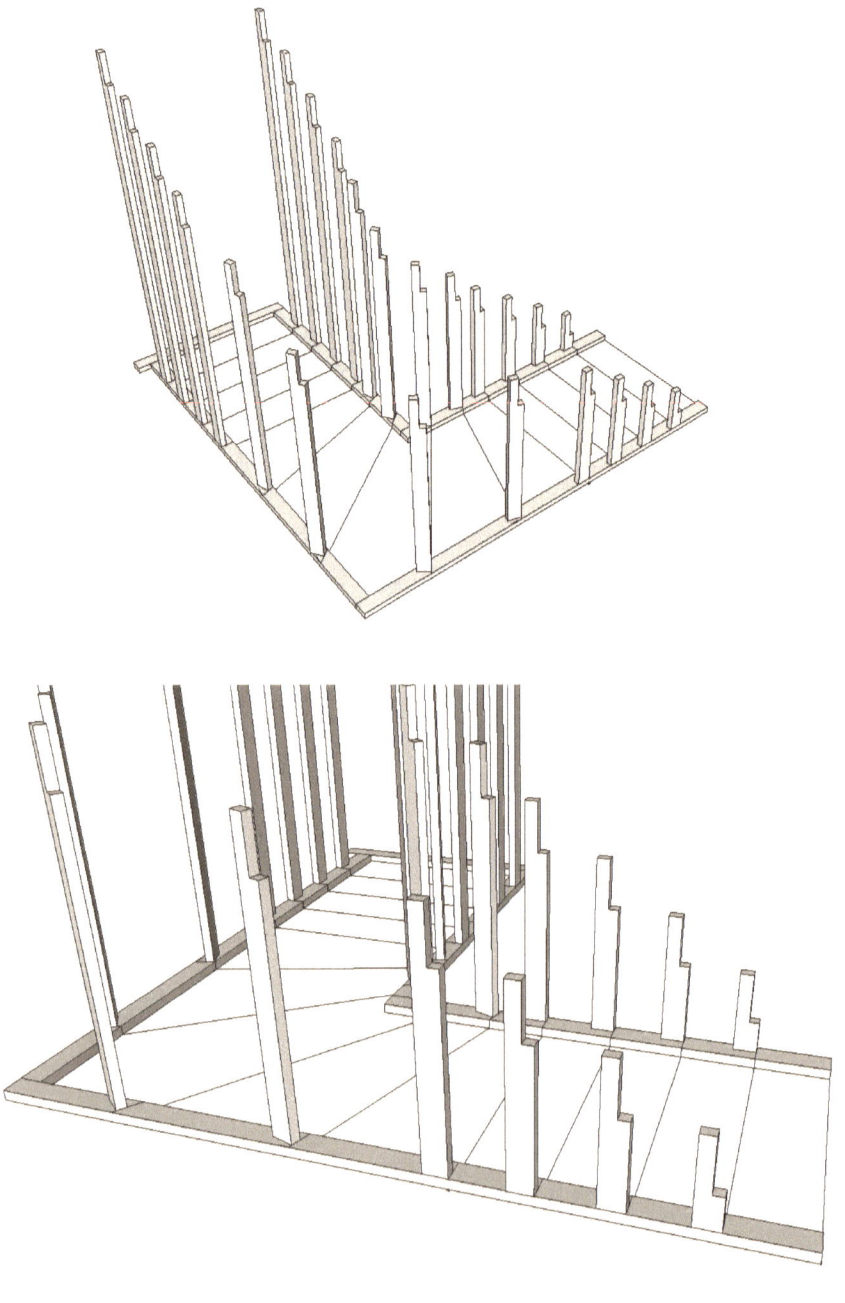

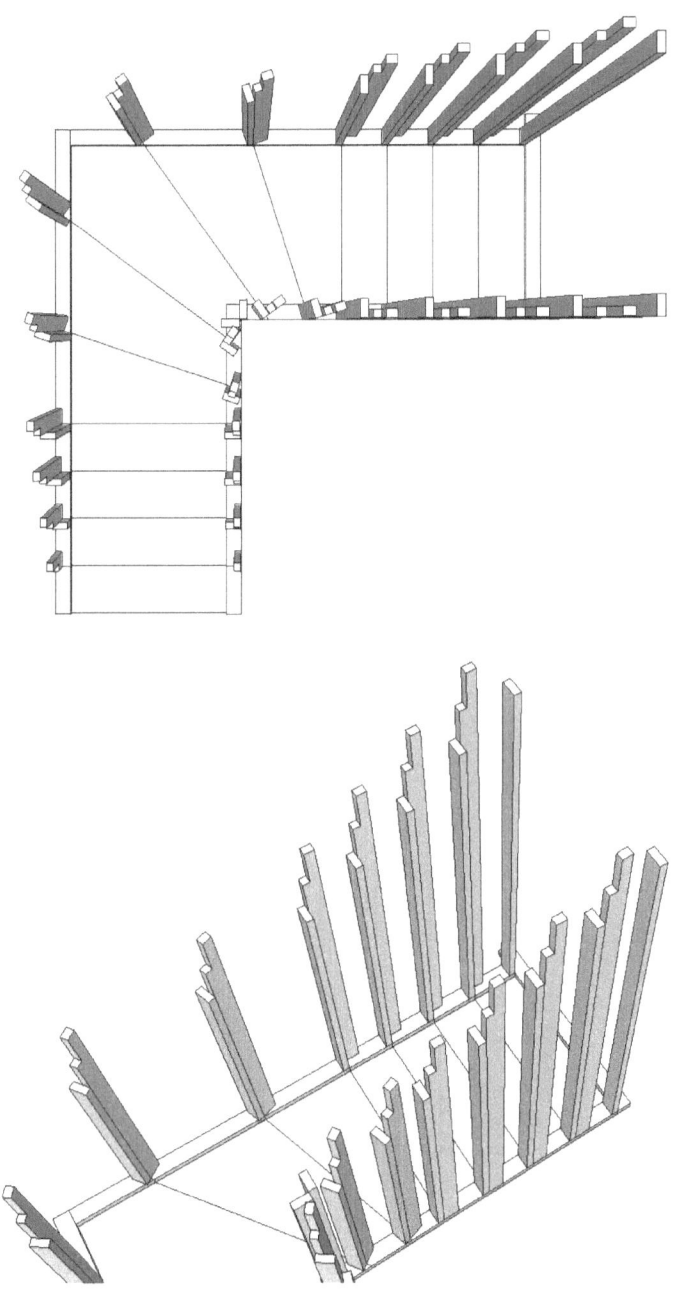

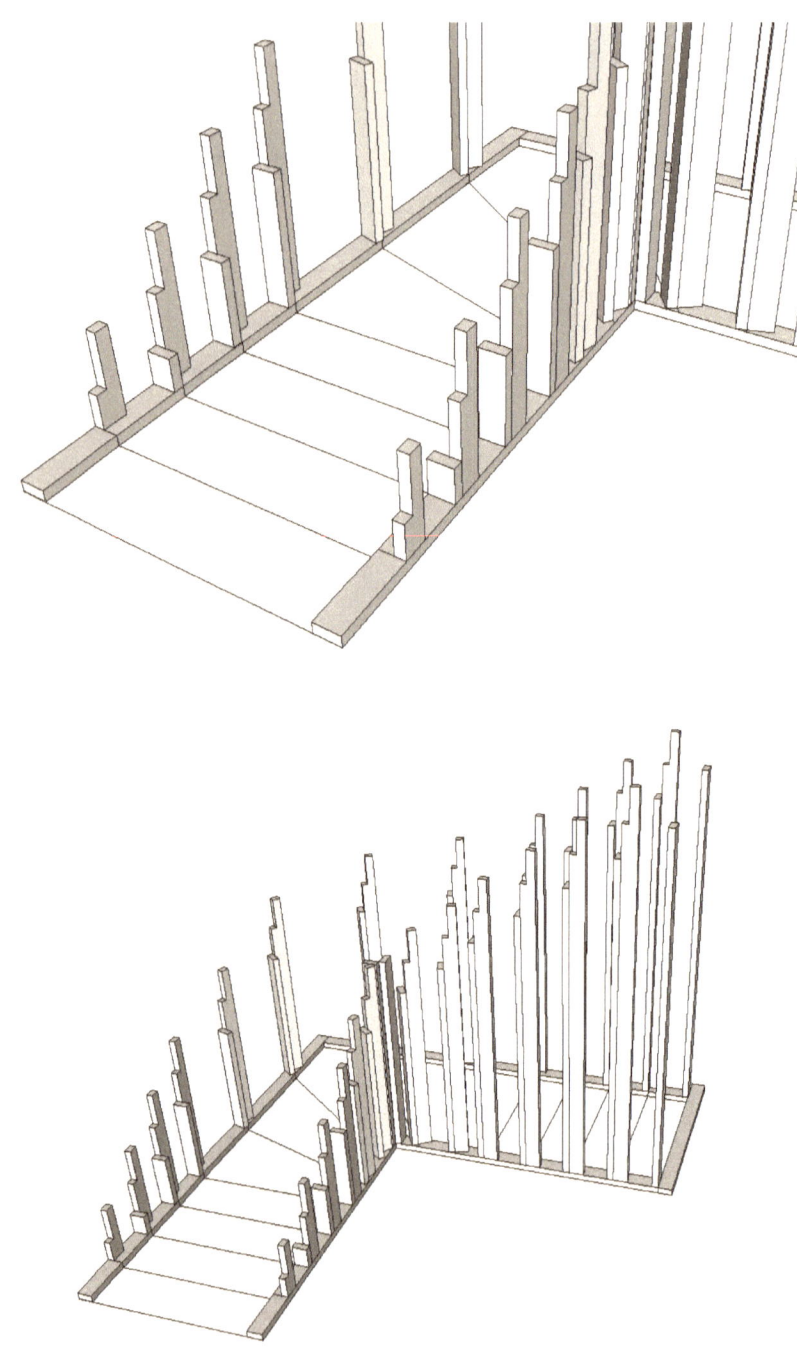

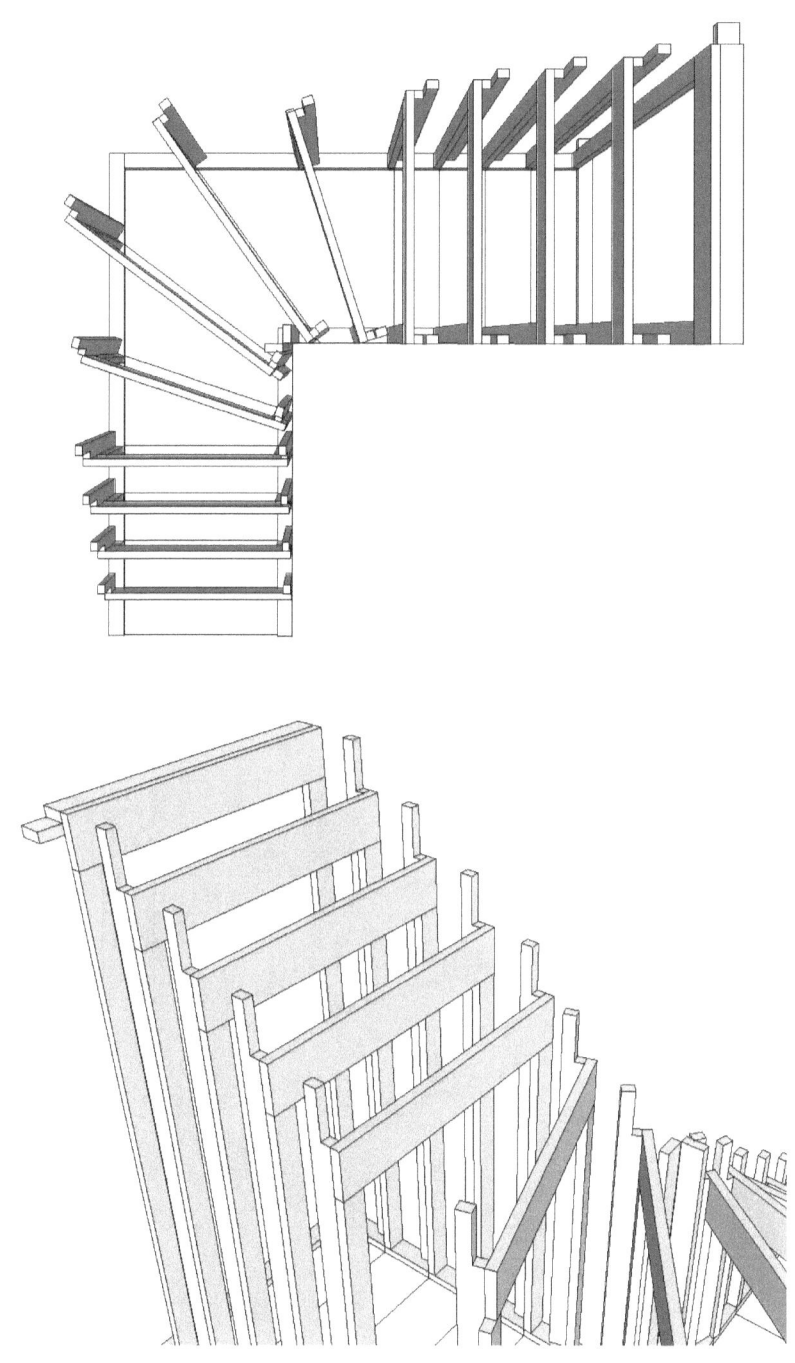

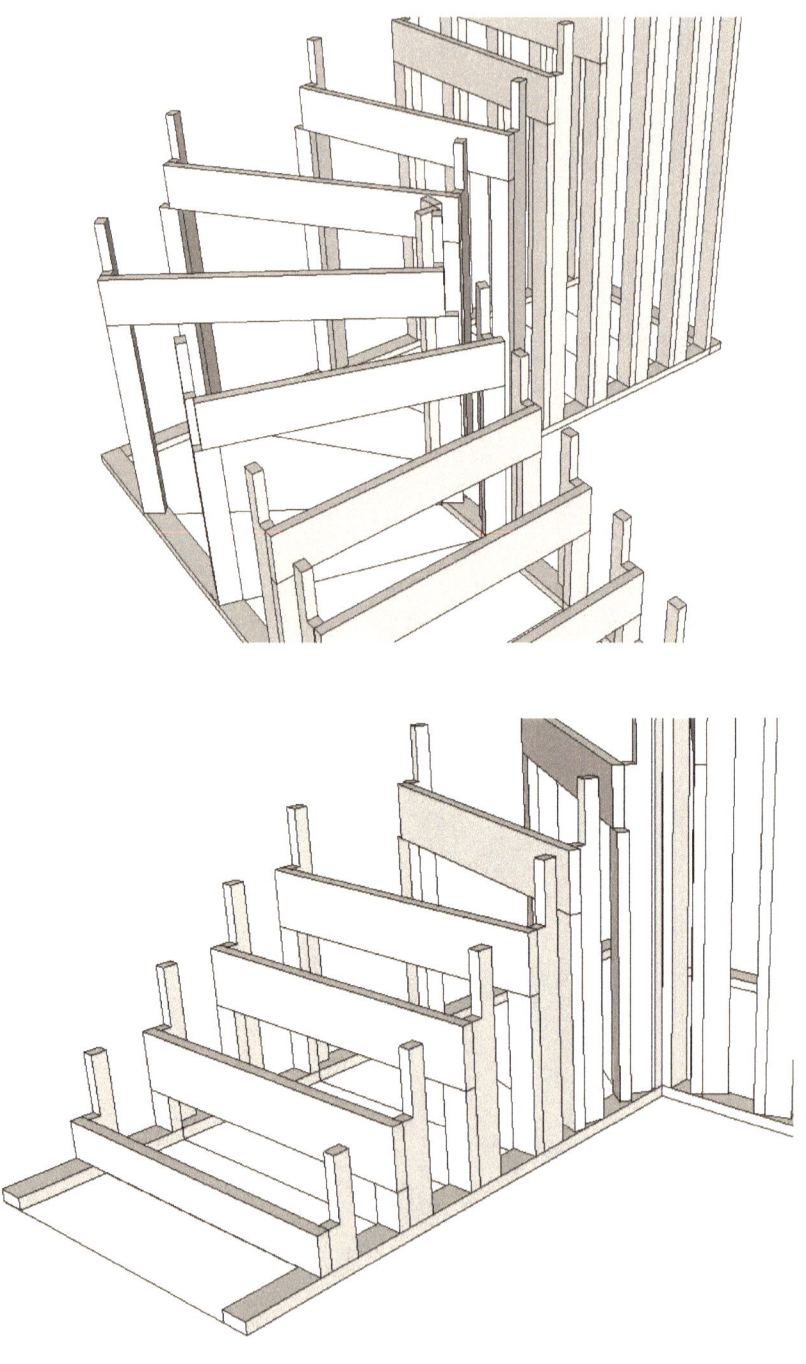

How to Layout Winder Treads

In this chapter I'm going up provide you with two ways to layout your winder stair treads. The first one can be used after you have laid out the stairway, but before it's built. This method usually works great for steps without 90° angles, but doesn't need to be used on steps with at least two 90° angles.

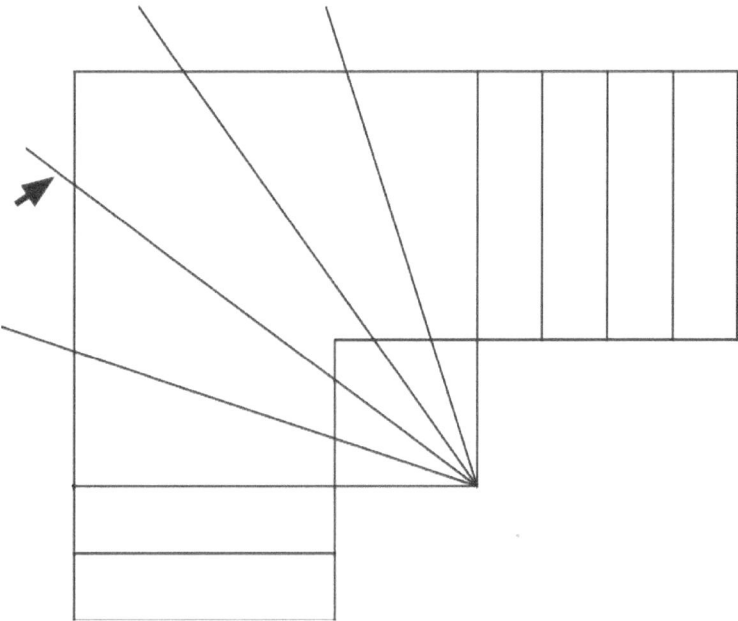

The first thing you need to do is extend the face of step lines to make it easier for you to see when marking pieces of plywood that will be used for the winder steps.

The next step will be to cut a few plywood pieces that are a little larger than each step. Then you can line the back of the scrap pieces up with the face of the next step as shown in next illustration.

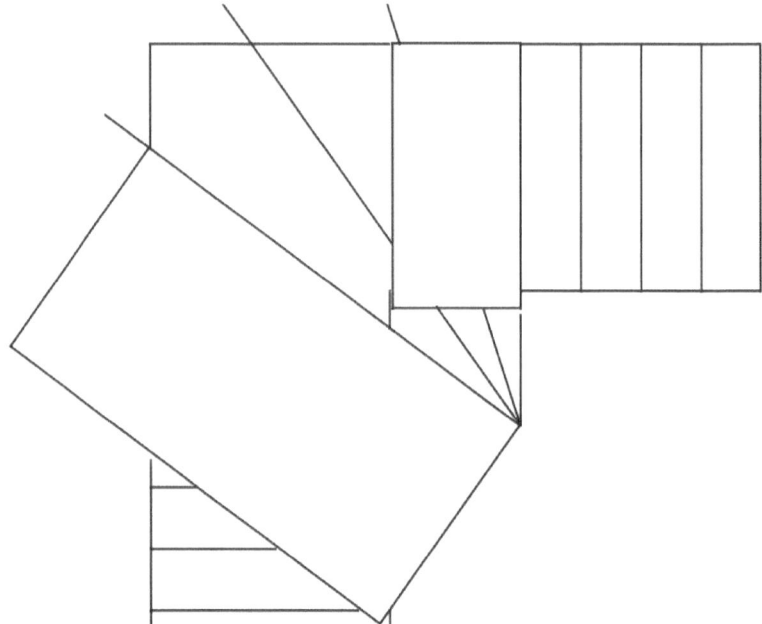

Then all you need to do is start connecting the dots. The arrows in the illustration below are pointing to the first lines that will be marked and then cut.

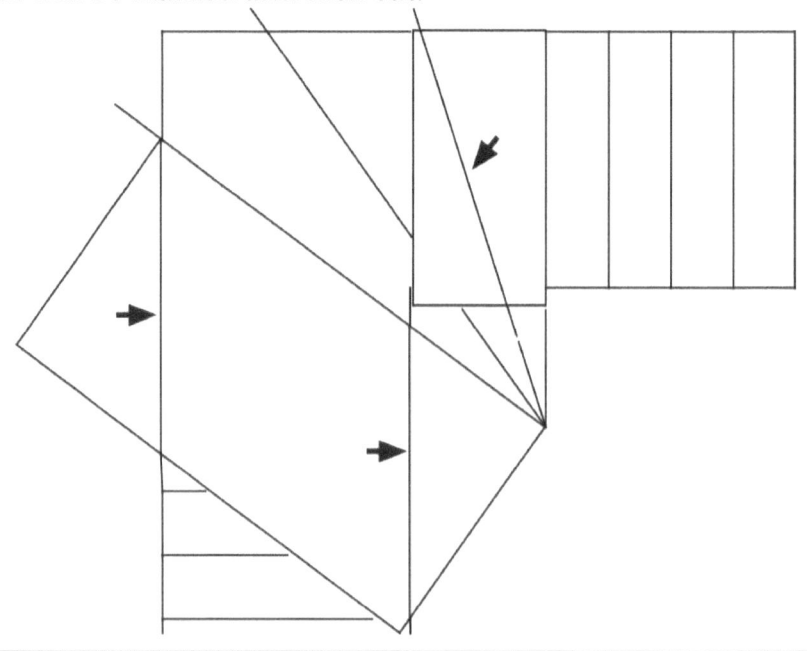

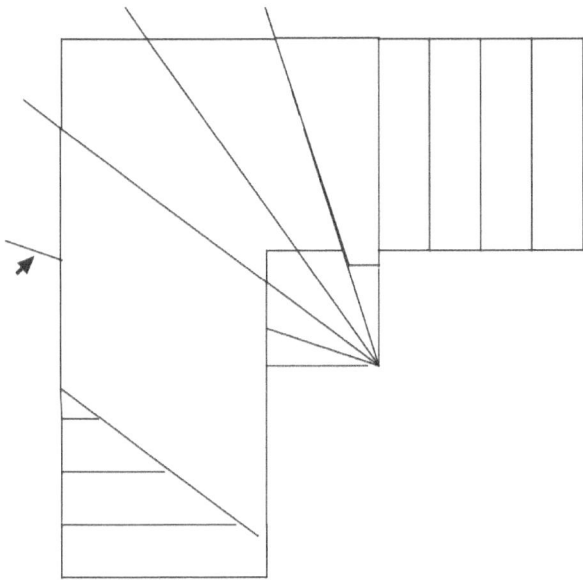

The extended face of step line is easier to see after we've cut the plywood. The arrows in the illustration below are pointing to the next lines that will need to be drawn and cut.

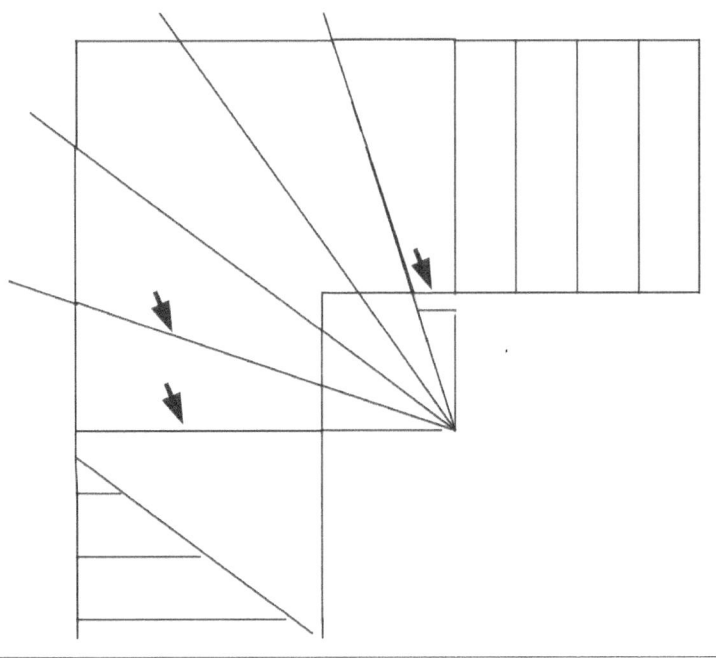

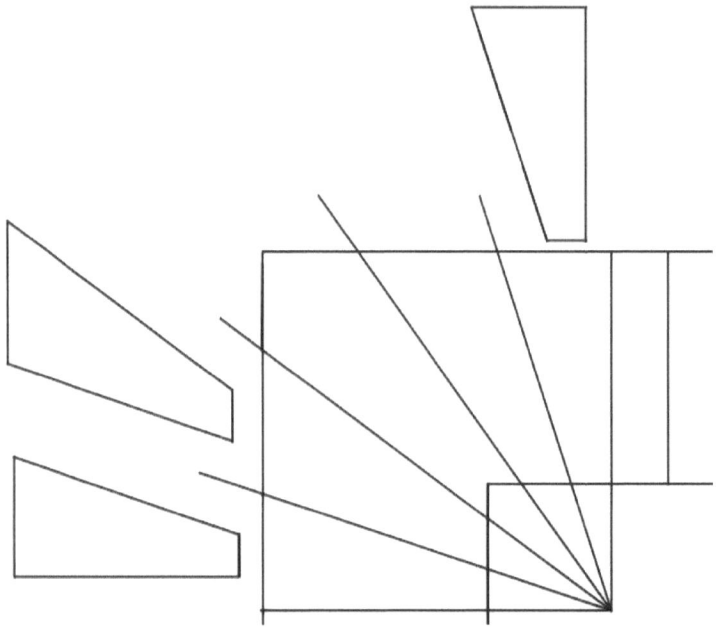

The illustration above provides us with the finished treads and below provides us with another method that can be used if there are at least two 90° angles per winder step.

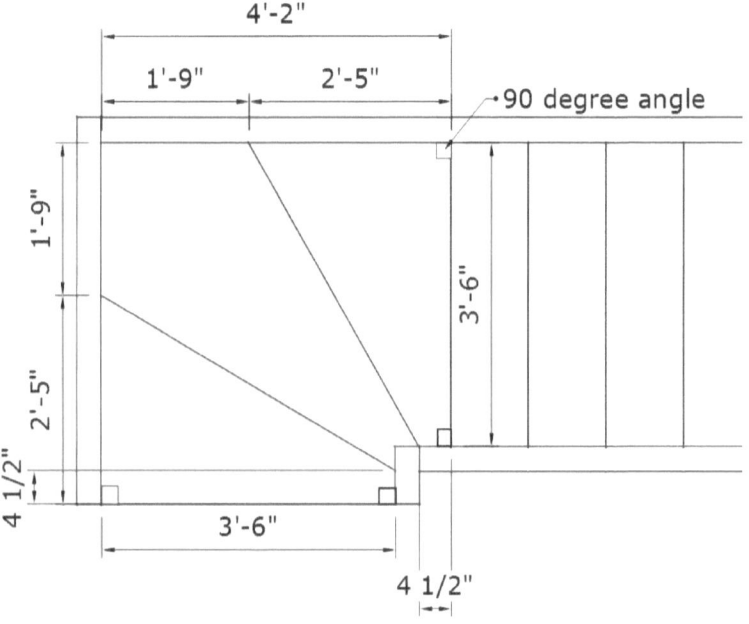

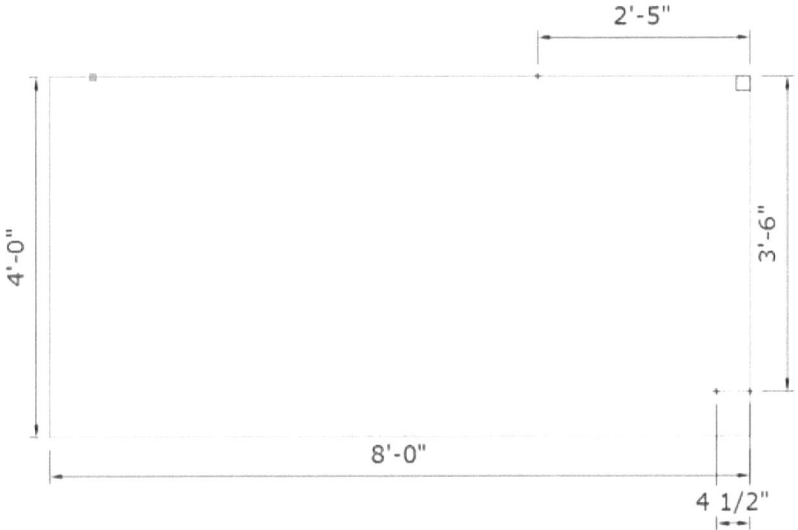

Using a sheet of 4 x 8 plywood, layout the winder step or tread using the measurements with 90° angles used in the floor plan and simply connect the dots as shown in illustration below.

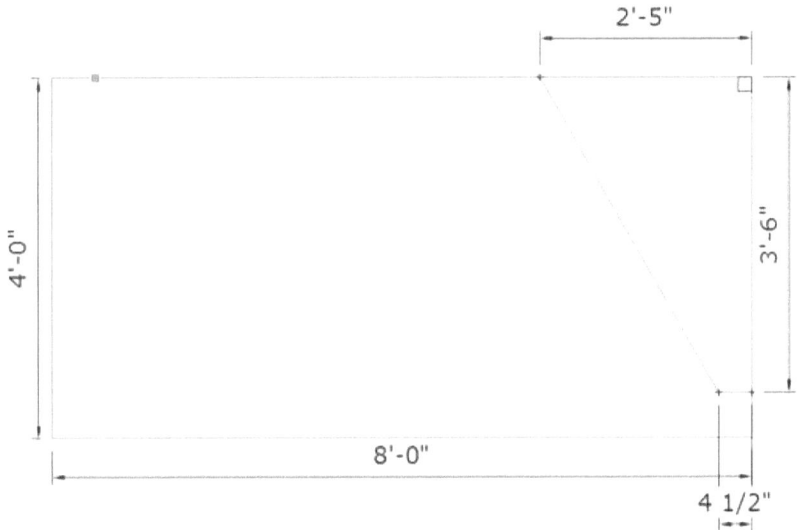

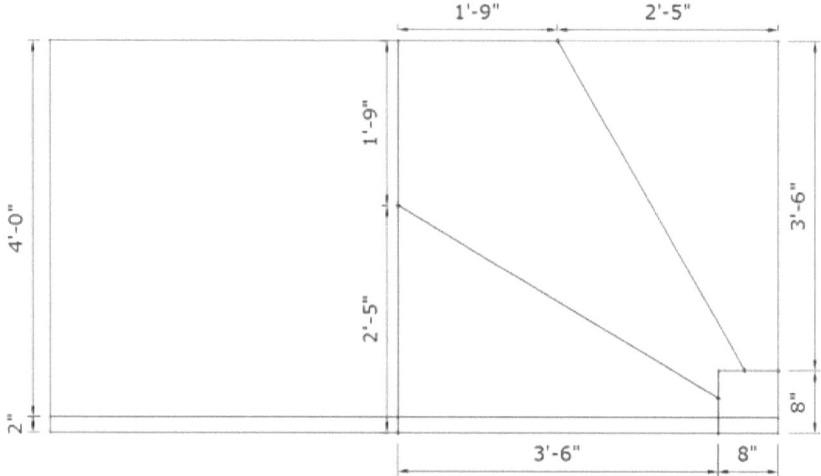

Here's another method that might be helpful and would work on winder steps with larger measurements.

Since the overall measurement of our winder square, for this example is 4'2", all we need to do is add a two inch strip to help us layout the winder steps.

We are only going to use it to layout the steps, but we don't need to use the piece with the 2 inch strip. On larger stairways with wider strips you could and it would make sense, but on smaller ones like this it usually won't.

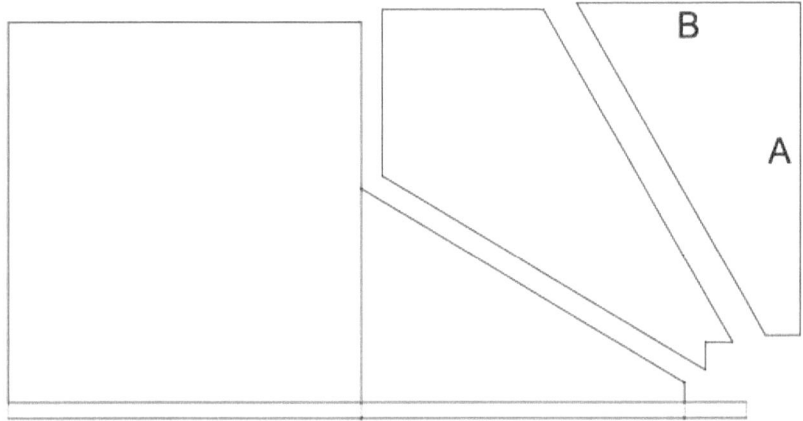

The first and last winder steps are going to be the same, except they will flip over when installed and you can use them as a pattern. The same would apply to the five step winder. The first and last will be the same and the second and fourth will be the same.

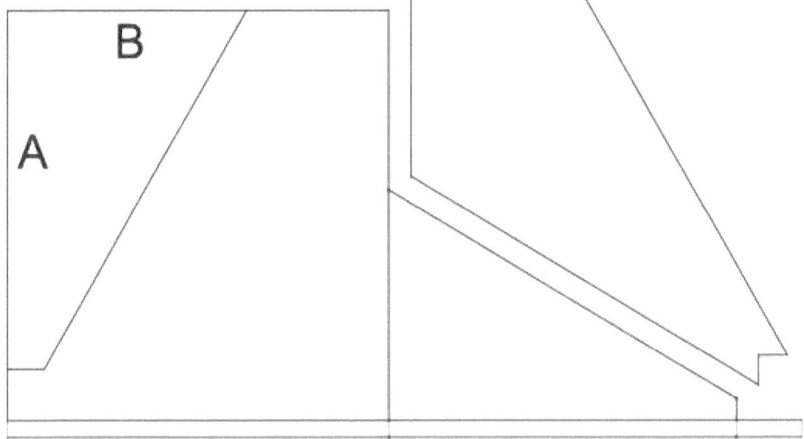

Build it in Sections

There are three ways you can build this type of stairway. Start from the top and work your way to the bottom, start from the bottom and work your way to the top or start from the angled winder steps located in the middle and then work your way toward the top and then toward the bottom.

I prefer working from the top down, but I strongly suggest using the method that will work best for you. It might be easier for you to calculate, layout and build the winder section first and then attach stringers or build a section of the stairway with the stringers and then assemble the winder steps.

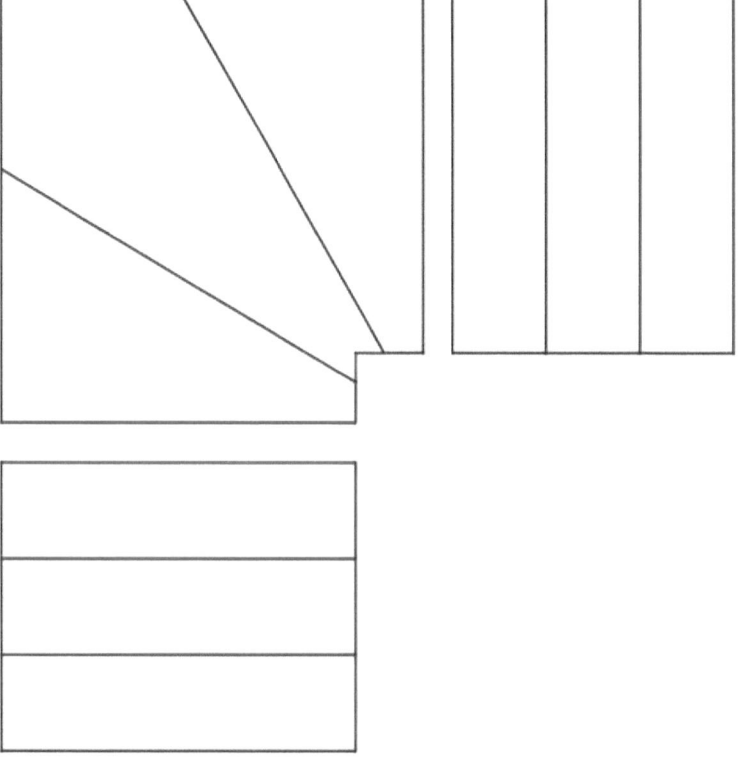

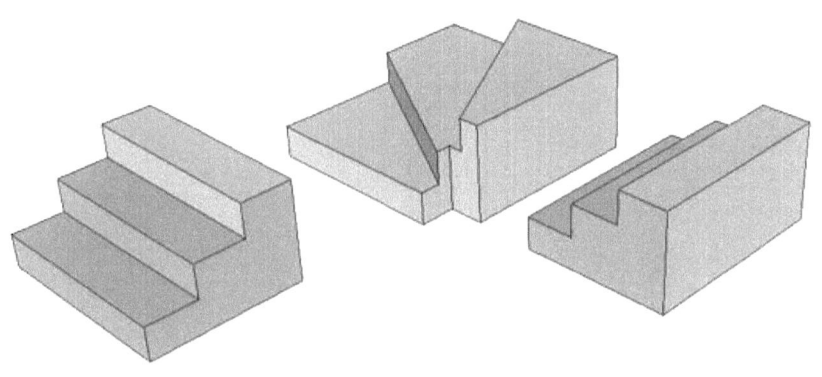
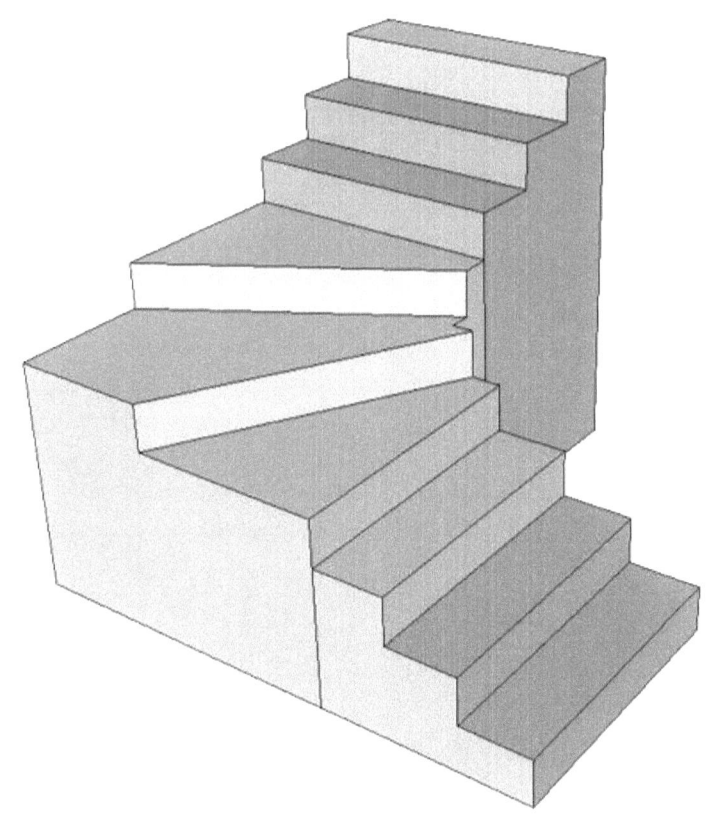

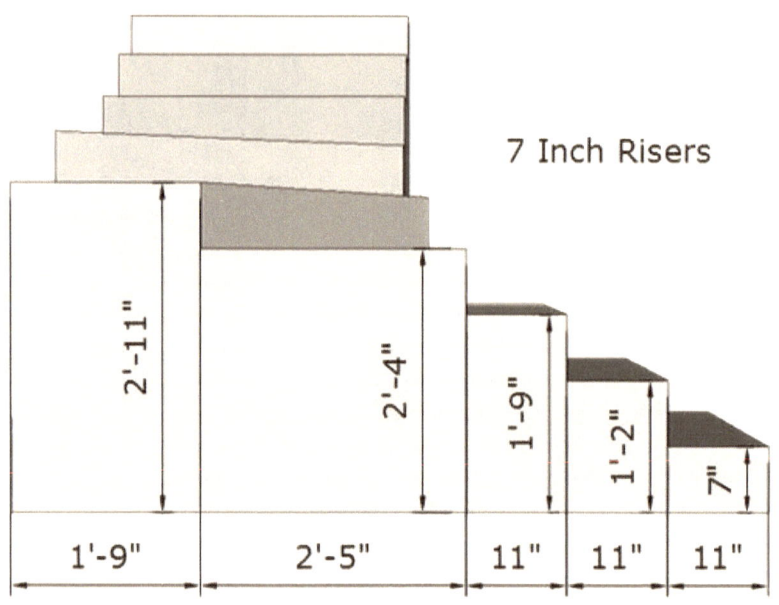

7 Inch Risers

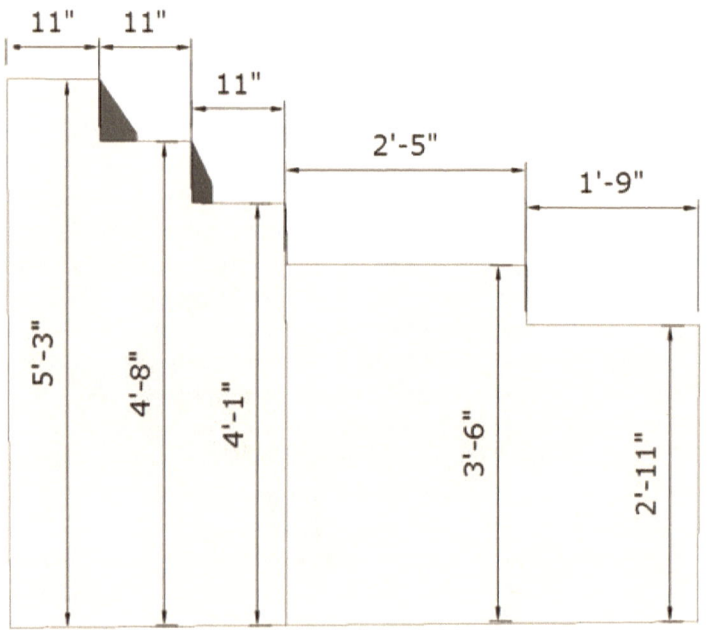

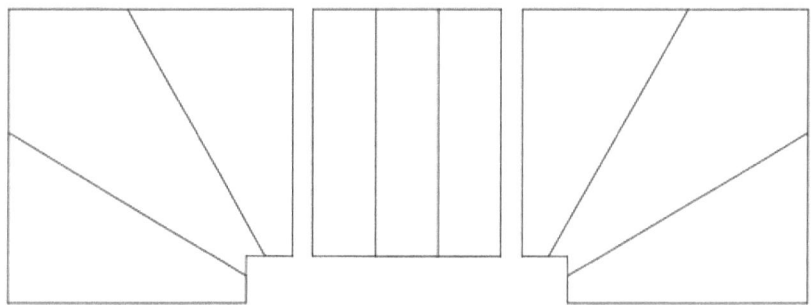

In this example we placed winder steps at both ends to give you another example, but the winder steps can be located using this section method of assembly anywhere in the stairway.

The example above has three angled winder steps (bottom section), then three rectangular shaped steps (middle section) and then three angled winder steps (upper section).

This method can be used regardless of the angled winder steps location in the stairway or whether or not you're going to use three angled winder steps or ten.

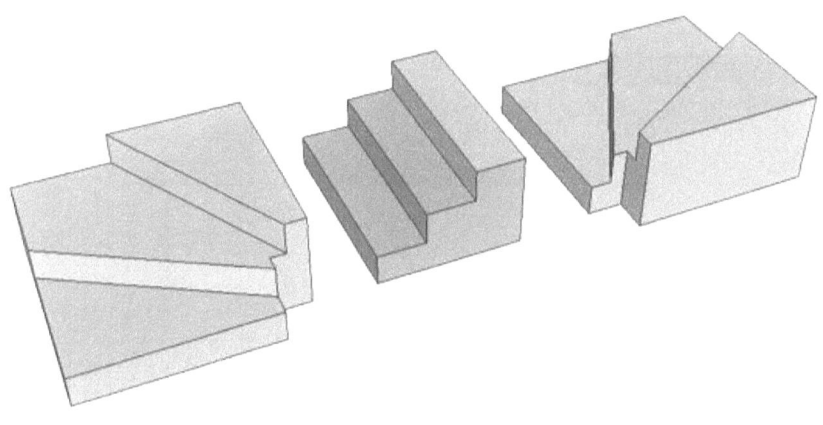

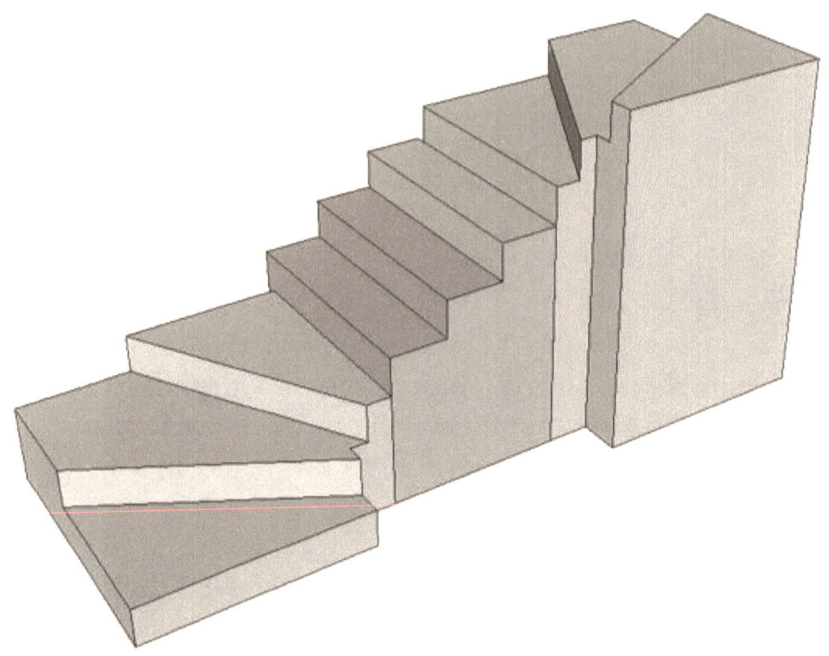

These two examples use 7 inch risers and I placed the measurements in both examples to give you a better idea of what the finished stairway would look like.

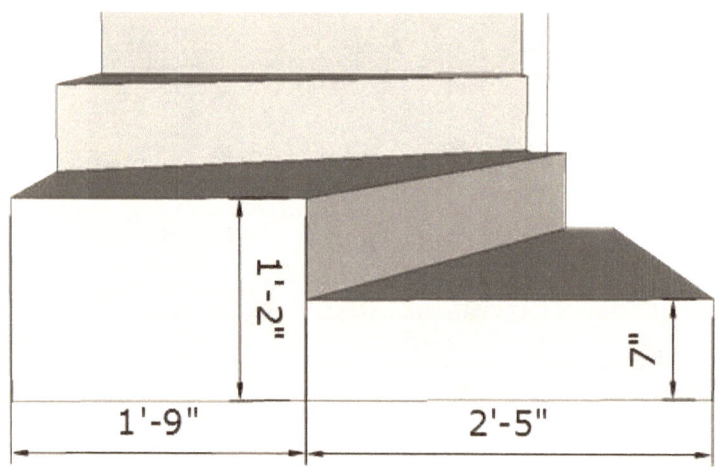

1'-9"
2'-5"
11"
11"
11"
2'-5"
1'-9"
4'-8"
4'-1"
3'-6"
2'-11"
2'-4"
1'-9"
1'-2"
4'-2"
2'-9"
4'-2"

2'-5"
1'-9"
5'-3"
4'-8"

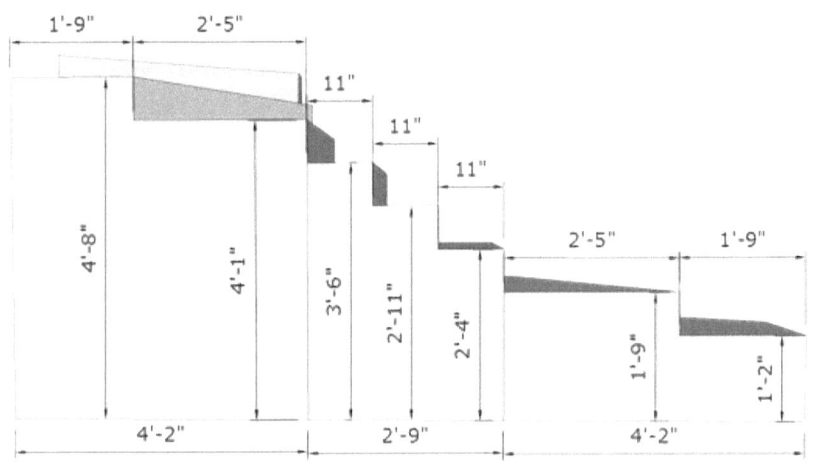

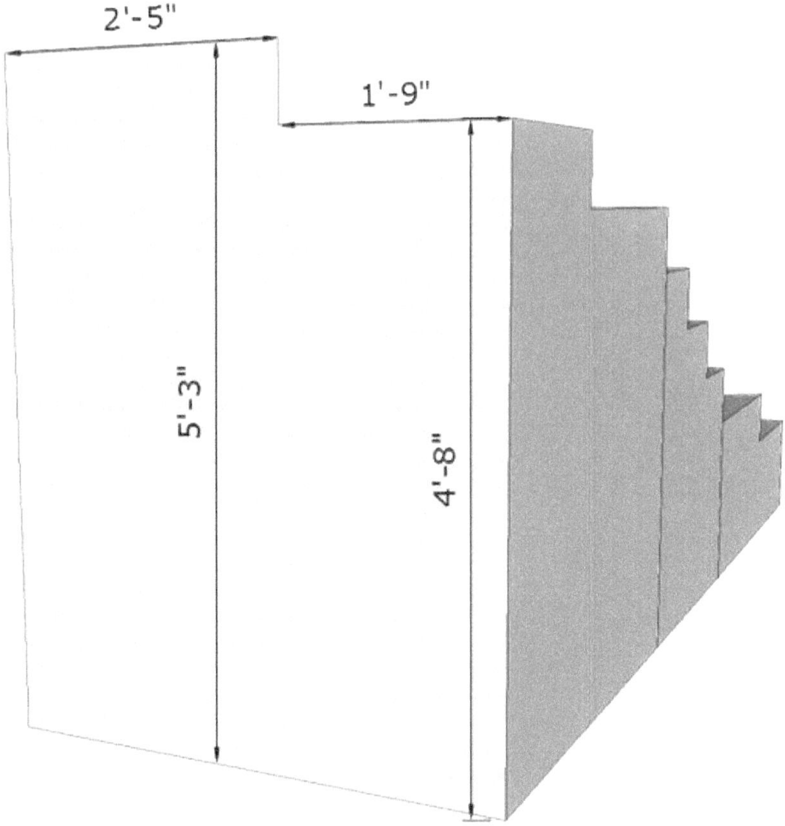

That's it for our examples and if you need more help you can always check out our videos or send me an email. The videos and contact information can be found on our website.

http://stairs4u.com

I would love to see a few pictures of your project when completed.

www.ingramcontent.com/pod-product-compliance
Lightning Source LLC
Chambersburg PA
CBHW030754180526
45163CB00003B/1017